SpringerBriefs in Finance

CW00557879

For further volumes:
http://www.springer.com/series/10282

Yasmine Hayek Kobeissi

Multifractal Financial Markets

An Alternative Approach to Asset and Risk Management

 Springer

Yasmine Hayek Kobeissi
44 Lillyville Road
London SW6 5DW
United Kingdom

ISSN 2193-1720   ISSN 2193-1739 (electronic)
ISBN 978-1-4614-4489-3   ISBN 978-1-4614-4490-9   (eBook)
DOI 10.1007/978-1-4614-4490-9
Springer New York Heidelberg Dordrecht London

Library of Congress Control Number: 201294625

Printed on acid-free paper

Springer is part of Springer Science+Business Media (www.springer.com)

To My Father

Preface: Pursuit of Certainty is Vain and Risky

En hommage à Benoit Mandelbrot (1924–2010)

The standard approach of the efficient markets concept dangerously underestimates the risks in the financial system. The fall of Long-Term Capital Management (LTCM) in 1998 and the CDO market "domino effect" in 2008 are some of many examples of the disastrous effects of market risk underestimation. The problem with existing financial models is that mostly they are based on the conventional Brownian model wherein contradictory evidence is worked around rather than explained and built upon. The efficient market theory promised "expected" returns through simple risk assessment and it utilized profile models based on risk aversion—models that were adopted by financial markets, economics department, and business schools who taught this to students as the basis of modern portfolio theory (MPT). In the 1970 and 1980s, according to Professor B, Mandelbrot (1924–2010), this approach became

> ...the guiding principle for many of the standard tools of modern finance ... taught in business schools and shrink-wrapped into financial software packages... a house built on sand. (Mandelbrot, The (Mis)Behaviour of Markets—A Fractal View of Risk, Ruin and Reward 2008).

Professor B. Mandelbrot changed the way that we view the world by introducing the fractal geometry of nature. He gave us a tool to describe different systems with common characteristics. Between 1960 and 1965, Mandelbrot introduced into economics a special form of the notion of invariance that later became essential in physics and took the term "scaling." In 1972, he introduced multifractals by replacing scaling by multiscaling. Throughout, he showed that scaling and multiscaling have many practical consequences. Applied to finance it allows us to understand market instability. Wild prices, fat tails (that is, heavy-tailed market distributions that exhibit extreme skewness or kurtosis), and the long-term memory effect, led Mandelbrot to view the financial series as a fractal series. Fractals have helped model the weather, measure online traffic, compress computer files, and analyze seismic tremors and the distributions of galaxies.

Mandelbrot's hope was to build a stronger financial industry by improving its systems of control and regulation. He developed a multifractal model with variable market times, exponential price distributions, and fractal generators. Some say that Mandelbrot described the financial system without explaining it, but can anyone elucidate it? He was a wise observer and had a special way of scrutinizing objects and nature. His hope was that people would accept the reality of risky markets and stop pretending otherwise... thinking in a fractal way in finance is about observing, analyzing, and trading opportunities arising from the markets anomalies!

By identifying markets' structures, what we essentially learn from Mandelbrot is that a measure of risk should take into account long-term price dependence and the tendency of bad news to arrive in waves. Our focus should not be on how to predict prices; but on how to foresee risks:

> Opportunities are in small packages of time, large price changes tend to cluster and follow one another. If there was a large price change yesterday, then today is a risky day. (Mandelbrot, The (Mis)Behaviour of Markets—A Fractal View of Risk, Ruin and Reward, 2008).

This is a delicate statement, meaning that we may not be able to forecast market direction, but during its clustery periods we either exit the market and thereby reduce the chances of loss or trade its volatility.

The question he left unanswered is how to do this in practice and the aim of this book is to seek an answer to that question. The book seeks to guide financial practitioners through the fractal approach and to help them develop appropriate models for estimating risks. Each cycle, each market, each strategy and each underlying asset requires an appropriate dedicated model. Nothing can be generalized to an entire portfolio. To analyze a market's dissimilarities is difficult and tricky. Each field needs its own appropriate—in the zone—skilled person. Ultimately, the book is dedicated to all those who do not want to fall in the next financial crisis and to those who want to get to the next level to perceive opportunities in markets turbulences.

I hope that this book will allow readers to discern the fractal nature of the systems around us.

Acknowledgments

I had the honor of working with Professor Mandelbrot and wish to acknowledge our correspondence on the various issues and comments offered by him.

Being Pr. Shefrin student although for a short period has helped me develop deeper understanding of market behavior especially through his challenging experiments.

I owe a lot Fred Viole and Professor David. Nawrocki for their comments on earlier versions, specifically the bifurcation entropy process and the subjective utility function and I am thus privileged to know their work intimately.

Special thanks also to Professor Martin Neil and Professor Norman Fenton for their time sharing with me their thoughts on how to apply Bayes theorem under conditions of uncertainty.

Several other people contributed in the many prior discussions in which I was forcibly educated. These include Markus Nordberg and Edgar Peters.

Mostly, I wish to thank all of the experts, researchers, authors listed, and not listed in the current bibliography without which progress cannot be made in the field.

My deepest gratitude goes to my husband Nabil Kobeissi for his macro economic analysis and constructive criticism during the preparation of the book.

I cannot thank enough Wayne Landing for his time and consideration editing and fine-tuning my ideas, he provided fantastic insights in how to put the book together.

I thank my little ones, Rayyan and Sarah, whose understanding and encouragement (continuously peeking on me to make sure I was advancing) made this book possible. The fractal brain branching image Fig. 1.3 is my Sarah's amazing way of helping me.

The views expressed in this book and any errors are the responsibility of the author.

Introduction

The field of portfolio management is constantly evolving, as new information technologies, economies, markets, and instruments continue to emerge, prompting it to adapt to the needs of the global financial environment in which it operates. Demands on a portfolio manager's performance are greater than ever and outperforming the market (delivering the "alpha") has increased the expectations of portfolio managers and investors alike. A portfolio's performance depends on the portfolio manager's abilities to make timely forecasts of trends and to foresee events in the financial markets. There are numerous ways of calculating market valuations and deciding on the most appropriate method to use requires a good understanding of the characteristics of financial markets.

The current body of financial literature offers numerous, diverse theories on how markets function, including the search for determinism, unpredictability, mathematical formalism and empiricism. However, no consensus has yet been reached on which approach most accurately portrays reality. The orthodox models as referred to by Mandelbrot do not accurately forecast market behavior as they are mere simplifications of a much more complex reality.

Actually, our focus should not be on how to predict prices but on how to foresee risks, but future risk evaluation requires projections on the roughness of prices/returns and fractal geometry is all about the study of roughness.

Fractals are a pattern or object whose parts echo the whole in a smaller scale. By spotting the symmetries or invariances, we are able to simplify things more efficiently and as such, identify the fundamental properties that remain constant from one object to another. Each small fractal segment contains the definition of the construction of the whole, characterized by a power-law dynamic. An example of a fractal is the lung, the structure of which consists of an infinite number of levels.

Mandelbrot (Mandelbrot 2008, p. 32), a mathematical scientist who studied financial charts as geometric objects, defined three stages of indeterminism: mild, slow, and wild randomness. Mild randomness is associated with Brownian behavior; slow randomness with log normality; and wild randomness with two distinct forms of wild behavior: the "Joseph effect" and "Noah effect". The Joseph effect refers to the persistence found in the financial system, while the Noah effect refers to the strong discontinuities in the system. Wild prices, fat tails, and the long-term memory effect led Mandelbrot to view the financial series as a fractal series.

Financial markets continuously evolve in a nonlinear, dynamic fashion, affected by destabilizing events that occur in the world and auto-reinforced mechanisms or memory effects. In spite of its complexities, a financial market is not entirely incomprehensible; although market cycles continuously change, the underlying mechanisms remain the same. In this book the paradigm of "Multifractal Markets", which views traditional approaches as subsets of a more global approach, is explained.

How does understanding the multifractal nature of the financial markets help us to better manage our portfolios?

Understanding the nature of the market allows us to use the cluster property of volatility as a tool to measure and forecast risk: A measure of risk that takes into account long-term price dependence and the tendency of bad news to cluster. The fractality of the system, because of its self-affine characteristic, allows cycles to be identified; these limited, non-periodic cycles depend on the processes at play and on the behavior of the market players. In reality financial systems appear to be generated by dynamic systems that are both deterministic and unpredictable. Determinism permits the majority of investors to identify opportunities while Unpredictability ensures the survival of markets and allows skilled managers to forecast opportunities.

The evolution of prices results from the dynamic relationship between the underlying mechanisms of the economy and financial markets and the biases of financial agents. Together these concepts allow us to understand a market in its different phases. The unpredictable and fractal nature of markets and the subjectivity of interpretations explain the economic and financial complexity.

This book consists of three main parts.

Part one looks at the multifractal market hypothesis. Determinism allows for the identification of short- and medium-term memory effects. Short-term price forecasting is possible if we can identify a structure of correlations between prices that is a Lévy stable process at play. However, beyond a certain period of time the system acquires a degree of freedom; although it keeps its fractal structure and remains confined to its attractor (the macro fundamentals). Unpredictability comes from the sensitivity of the system to initial conditions and the detection of a process that evolves parallel to the Lévy-stable process. While it is true that exogenous irregularities are partly responsible for the disturbances in systems, the largest disturbances have their origin in the intrinsic element of financial systems, mainly from information that has not been disseminated, correlated investment horizons and excessive leverage. Here, we introduce the concept of "noisy chaos" which takes into account the sensitivities of the financial systems based on the processes of bifurcation, entropy named as (BE), and convergence that occur at the heart of the instability in the financial markets. The key here is to identify and measure indicators to be able to construct a model that accounts for extreme consensus in estimating market reversals or "switching points". At this stage, price forecasting is impossible; risk forecasting becomes our only tool against markets swings.

Part two analyses the main concepts of perception in order to understand how biases made at any stage of the decision-making process can lead to (mis)behavior and consequently to market instability. The sense of time is different for every investor. For short-term traders, an hour or a day can be a full cycle but for long-term investors a full cycle may be 3 years or more. Short-term investors perceive the risk and return of an asset in "intrinsic time", that is, a measure that counts the number of trading opportunities, regardless of the calendar time that passes between them. Some investors have learned how to react to market information without too many emotions interfering in their decision-making process. Some are so fast in reacting and in interconnecting facts that we can compare them to drummers, whose thoughts are always ahead of the beat and this is where frequency traders find their edge. Here, we show how even though investor behavior is highly variable and diverse, it is still possible to categorize them according to heuristics they relate to under uncertainty.

As new information is constantly entering the market financial participants need to revise their expectations according to their own utility perception. As such the study of utility is important to understand the financial marketplace. The key element in any information content is the surprise element. Surprise is experienced only if an unexpected outcome occurs from which a new or different utility per individual is derived. Bearing in mind that information is a decreasing function of probability, we introduce an innovative subjective utility theory as per the findings of Viole and Nawrocki: "Multiple Heterogeneous Benchmark Utility Functions". Bayes theorem and fuzzy logic, that have found applications in many contexts, are presented as a device to effectively account for "probabilities" in the decision making under uncertainty.

In part three, macrostates indicators previously explained are explored. Economic cycles are the strange attractor of the financial process. Economic indicators and their interrelationships with each other are reviewed in Chap. 4. The specifics of each cycle make it unique; economic indicators by themselves are not as valuable, but rather their causes, and magnitudes effects. With this interconnectedness in mind, we become able to explore ways to perceive the cycle ahead. Chapter 5 introduces dynamic management approaches for trading in multifractal financial markets, highlighting their advantages, applications, and limitations and also illustrates the importance of macro and tail risk management in identifying effective strategies. The art of successful tail risk management lies in the ability to hedge against sudden market drifts or any specific micro market risk as well as against long periods of low volatility. It is also about attempting to time the volatility and benefit from its characteristics without having to rely on seismic events to gain profits.

Avoiding risk should be what investors and portfolio managers care about the most, although when investing their first thought is usually how much they can gain, their main concern is at what cost, but unfortunately they are then inevitably swayed by market biases. It is important to think in terms of affordable risks before thinking of potential gains. Forecasting might require sophisticated mathematical, economic, and financial tools in order to account for variable and uncertain factors such as people's hopes and fears, natural catastrophes and new inventions. But most of all it requires a sophisticated mind to grasp a market's structure and humility to accept it.

Throughout the book, empirical evidence for finance market instability is given, together with a history of the explosion of the "tech bubble" in 2001 and the housing market collapse in 2008 to ensure that practical considerations have not been overlooked. Research and most intriguingly in-the-lab findings that we came across are presented as well as detailed equity valuation methods. Although the book includes mathematical explanations, the writing technique is on a level aimed at widening its appeal to the broader financial market as well as to graduate level students of finance and economics. It is written in a way that allows the reader to be aware of the fractal nature of the financial markets without having to solve mathematical equations. Mathematical details in Chaps. 1 and 2 are for readers who would like to locate references for further researches. Remember that fractal is the geometry of roughness. The aim is not to prove mathematically that financial series are fractal. It has been done by Professor Benoit Mandelbrot decades ago. The aim is to help you discern similarities in complex structures, to perceive financial markets from a different approach. You will be able to seek alpha according to market risks and your level of loss aversion i.e., how much money can you afford to lose (loss budget).

An offensive modeling approach cannot merely rely on statistical risk measures as it cannot capture all of the possible risks. The aim of the book is not to develop entirely new and or completely infallible forecasting tools, desirable as they may be, practically it would be impossible. Markets keep shifting and in doing so they break models. By accepting the reality of uncertain markets this book provides financial practitioners with the elements to develop enhanced portfolio management tools. The mystery of the financial markets is the key to its existence, understanding this is the key to our survival. Survival in multifractal markets requires us to mind the risk.

Figures, Diagrams and Tables

Contents

Chapter 1
Turbulence in the Financial Markets

Abstract Financial markets continuously evolve in a nonlinear, dynamic fashion, affected by destabilizing events and auto-reinforced mechanisms or memory effects. In spite of its complexities, the financial market is not entirely incomprehensible; although market cycles continuously change, their underlying mechanisms remain the same. This chapter provides an explanation of how fractal geometry helps us to understand the mechanisms underlying financial markets. One of the main features of financial markets is the alternation of periods of large price changes with periods of smaller changes. Fluctuations in volatility are unrelated to the predictability of future returns. This statement implies that there is autocorrelation structures dependence in the absolute values of returns. The multifractal model of asset returns combines the properties of L-stable processes (stationary and independent stable increments) and fractional Brownian Motions (tendency of price changes to be followed by changes in the same (or opposite) direction) to allow for long tails, correlated volatilities, and either unpredictability or long memory in returns.

In October 1987, the Dow Jones Industrial Average plunged more than 20 % in one day, an unlikely 20-standard deviation event whose probability of occurrence is less than one in ten to the 50th power (Mandelbrot 2008, p. 4). In September 2008, the Dow Jones once again dropped significantly, declining by more than 7 % in one day, a probability of 1 in 50 billion. Under conventional financial theory, these sharp drops in the stock market were not supposed to happen.

Y. Hayek Kobeissi, *Multifractal Financial Markets*, SpringerBriefs in Finance, DOI: 10.1007/978-1-4614-4490-9_1, © The Author(s) 2013

1.1 (Mis)Behavior of Markets

The efficient market theory[1] promised easy returns through simple risk valuation and profile models based on risk aversion. In this model, option markets flourished and structured finance boomed. The hypothesis of efficiency requires that certain conditions are met: the rationality and homogeneity of investors; free distribution of reliable information; liquidity of markets; and absence of transaction costs. In reality, however, these conditions do not hold—market participants perceive information differently from one another; transaction costs, especially spreads, are considerable; and liquidity issues are often present. These market anomalies cannot be overlooked as they are an integral part of the financial markets. The degree of risk aversion and risk seeking is highly variable as the interpretation and assimilation of information varies from one group of investors to another. That leads to the observed fluctuations in the stock prices.

1.1.1 Heterogeneous Assimilation of Information

The main determinant of the value of a stock is the expectation of future profits. The availability of information that affects this expectation is an important factor in the determination of its price. The degree to which available information affects a stock price depends on the ability of the majority of investors to proficiently interpret this information thus reducing the gap between the market price and the fundamental value. In practice, however, information is not necessarily immediately reflected in prices, it is neither instantaneously assimilated nor disseminated in the same manner. The existence of periods of optimism and pessimism are proof of this imperfect environment. Our time is marked by the large number of "e-traders" who access the markets by online means. The behavior and motivations of these participants are not homogenous and not always rational, which leads to an ever-changing financial market.

---

[1] Classical financial theory assumes that financial markets are efficient. A review of the efficient market hypothesis, however, shows that many of its required conditions are unachievable in reality. First introduced by Louis Bachelier in the 1900s, the concept of an efficient market assumes that competition among a large number of rational investors eventually lead to equilibrium and the resulting equilibrium prices reflect the information content of past or anticipated events. In other words, equilibrium stock prices must be, in principle, equal to their fundamental value and deviations from these equilibrium prices reflect the level of uncertainty at any given moment. As such, the principles of financial markets are identical to those of a roulette game where the players do not have memory or if they do, is too short to be able to take past experiences into consideration. The efficient market approach affirms the unpredictable nature of the financial markets; yields are independent and prices follow a random pattern, that is, a Brownian motion. This randomness obeys a law of probability often attributed to the Laplace Gauss Law of normal distribution (Hayek 2010).

1.1.2   Correlated Investments Horizons Time

Information is perceived according to the investment horizon of each investor. Different time horizons assure liquidity. E. Peters (Peters 1996) asserts that financial markets series are characterized by a long but finite memory, the duration (N) of the memory varies from one system to another, as well as from one stock to another. The auto-regressive process works well in the short- and the medium-term; and the observed prices reflect a combination of short- and medium-term trading and long-term fundamental analysis.

Often, investors are incapable of analyzing certain information on their own and as such, rely on professional investors to make up their minds. Then again, as professionals are primarily guided by "benchmark tracking", that is, the constraints placed on the funds that they manage to generate the "alpha", they also follow market consensus to avoid being marginalized. These and other behavioral biases can destroy the stability of the market. The impending problems start when long-horizon investors actually behave as short-horizon investors draining liquidity from the long side.[2] Risk is related to liquidity and time and more specifically to diversifications in investments horizons. A long-term investor can see opportunities when a short-term trader drops a value stock, hence insuring the survival of the stock. When the long-term investors does not step in, then the market becomes illiquid triggering the process of disengagement leading to market instability. For instance, "Short-termism" refers to an excessive focus on short-term results at the expense of long-term interests (sustainable investments for the company): An excessive short-term focus by some corporate leaders, investors, and analysts (by focusing on the Earnings per share (EPS) only) combined with lack of attention to the fundamentals of long value creation and investments.[3] At the arrival of information, investors react according to their individual particular expectation based on their investment horizon. This notion is developed in the following chapters.

1.1.3   Cyclical Markets, Economies, and Human Nature

A financial market defines its own rules, which it continues to obey as long as no new information arrives that undermine or render obsolete those rules. For example, during periods of economic expansion, the stock market behavior is characterized as follows:

---

[2] The inverse is good, as liquidity shifts to the long side, it is the end of the bear market, the bottom has been reached.

[3] John Graham, Campbell Harvey and Shivaram Rajgopal has shown that managers are making real decisions—such as decreasing spending on research and development, maintenance and hiring of critical employees—in order to hit quarterly earnings targets they have provided as part of their own guidance (Graham, November/December 2006).

- tax-motivated selling in December;
- rebounds in January;
- rally between February and May;
- retreat in June, which is a period of normal profit-taking after the rally and before the earnings announcement;
- Another rebound in July, during which good news is integrated into the prices and a wave of profit taking occurs as investors start to focus more on the fundamental values prior to the holiday season.

The value attributed to any asset relies on an estimate of its future return. Valuations are the result of the economic process itself, which reflects the waves of optimism and pessimism brought about by simple factors or events that affect human thoughts and actions. The hypothesis of efficient markets assumes the existence of the rational investor in its economic model. This assumption overlooks an essential and variable component of human behavior, that is, that human beings are not objective creatures or economic beings that act rationally.[4] Human behavior is never constant; it vacillates from one extreme to the other and varies from one cycle to another.

Wild prices, fat tails and the long-term memory effect, led Mandelbrot to view the financial series as a fractal series. In the following section, we explain the fractal patterns and fractal geometry underlying the multifractal market hypothesis.

1.2 Fractal Geometry in the Context of "Roughness"

Mandelbrot developed fractal geometry to deal with the rough, irregular, and jagged objects he called "fractals," which come from the Latin term for "broken".[5] Fractals are a pattern or object whose parts echo the whole in a smaller scale. The key to identifying a fractal is to spot the regular within the irregular, the pattern in the absence of form. By spotting the symmetries or invariants, we are able to simplify things more efficiently and as such, identify the fundamental properties that remain constant from one object to another. Based on these descriptions, straight lines and perfect circles are not fractal objects.

> "How long is the coast of Britain?" is a question made famous by Mandelbrot; he explained that we cannot measure the length of coasts owing to their structure's recursive schema; that even in the smallest of bays, we can find an even smaller one, or in harbors, other smaller harbors, such that the total length of the coast approaches infinity (Mandelbrot 1967).

---

[4] The human being becomes economic during the consolidation phases following a market correction. This behavior is explained by the fact that the variety of financial agents is reduced to professionals who have been re-oriented toward the fundamental financial and economic data.

[5] Mandelbrot initially presented a brief description of the multifractal concept, which he then expanded in 1975.

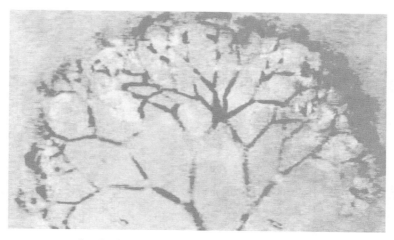

Our Brains are fractal with far reaching branches
Fractal brain branching by Sarah Kobeissi

Fig. 1.1 Fractal brain branching

1.2.1 Fractals

Fractal branching is seen in the lung, brain (Fig. 1.1), small intestine, blood vessels of the heart; some neurons, and in the DNA structure. Dr. Andras Pellionisz states that:

> Protein synthesis is not achieved by a masterstroke of "gene" information. It is an iterative process with recursive access to DNA information.

Fractal models are present in various contexts, including the turbulence in fluid dynamics, internet traffic, finance, image modeling, texture synthesis, meteorology, geophysics, and embryogenesis. Fractals are also found in ancient architecture around the world, demonstrating that structural repetition across several scales is common to many cultures. The following outlines the most important characteristics of fractals:

- Although fractals vary immensely, they also share common traits. For instance, they can all be scaled up or down by a specific amount.
- A Fractal is scaled the same manner in all directions, that is, it is self-similar (Fig. 1.2). For instance, a probability density function is statistically self-similar; when its statistical characteristics remain the same over time.
- Some fractals are self-affine and have multiple scaling factors put to play at once. These are called multifractals and may be scaled in many different ways at different points (Fig. 1.3).

Fig. 1.2 The Sierpinski
Gasket: A self-similar gasket
made of copies of itself.
Source Math 190a course
notes, Yale University, http://
classes.yale.edu/fractals/
IntroToFrac/SelfSim/
SelfSim.html

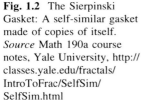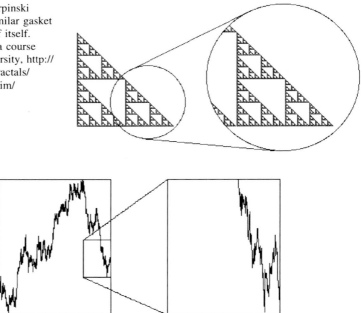

Fig. 1.3 Statistical self-similarity. *Source* Math 190a course notes, Yale University, http://
classes.yale.edu/fractals/

- All fractals start simply. The simplest fractals are constructed by iteration. A
  fractal starts with the "initiator" or the starting shape, followed by the "generator"
  or scaled copies of the initiator and then the rule of recursion.

   In certain deterministic fractals, like the Sierpinski Triangle, small pieces look
the same as the entire object. In random fractals, small increments of time will be
statistically similar to larger increments of time.

1.2.2 Fractal Geometry

Fractal geometry refers to the mathematics of fractals; it is a tool of analysis and
synthesis that recognizes, quantifies, and manipulates repeating patterns. These are
the main characteristics of fractal patterns:

- Pattern can take many forms; both concrete (the shape of a fern leaf) and
  abstract (shape of a probability). They can be scaled up or down and manipulated in various ways.
- The usage of patterns can be defined by deterministic rules or entirely left to
  chance.

- Each small fractal segment contains a definition of the construction of the whole, characterized by the power-law dynamic.

1.3 Multifractal Properties in Finance

In examining financial price series, we observe long periods of persistency as past information continuous to influence actual prices, followed by short periods of extreme variation or clusters. Instead of investing significant time and effort predicting the unpredictable, Mandelbrot decided to uncover the financial system process.

1.3.1 Long-Term Memory Dependence and Extreme Changes

The combination of long-term dependence and discontinuity makes the financial market unpredictable.

> The Joseph effect depends on the precise order of events, while the Noah effect depends on the relative size of each event (Mandelbrot 2008, p. 197).

Investors react to news in two major ways: over-reaction or under-reaction. Market prices do not always reflect all available information. Some information is stored/ignored and reactions come later. The stored information is only integrated with current knowledge after its validity or importance has been established. Old information also carries weight in the actual asset price valuation. Past information influences present valuation which then influences the future.

Significant changes tend to cluster and follow one another, although their directions are not necessarily the same. For instance, a huge fall today can be followed by either a similarly significant fall or rise tomorrow. Modern quantitative approaches to risk management and measurement view the distribution of returns as a log normal distribution, that is, the price variations follow the bell curve. Such distributions, however, underestimate the probability of extreme fluctuations in prices, which occur in time of panic. In reality, information arrives in a discontinuous manner and consequently, if the distribution of information clusters, the distribution of returns will most probably cluster as well. Sometimes, markets react immediately to unexpected events or shocks and that reaction is magnified disproportionally and exponentially by the different participants. This over-reaction of the market is the "Noah effect". Most investors fail to integrate old and new information properly; thus, their emotions lead to over-reactions in the returns cross-section and excess volatility in the return time series. In such situations, even the advantages of diversification are rendered obsolete because of the financial disengagement that is generalized for all financial products in all markets where correlations tend to one. Figure 1.4 illustrates the Noah effect during an intraday trading session.

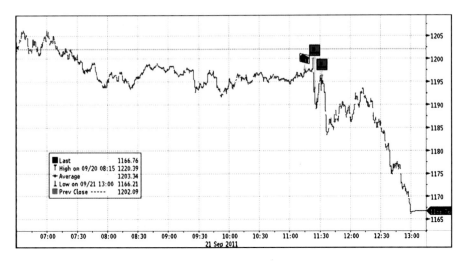

Fig. 1.4 The Noah effect (SPX Index). *Source* Bloomberg

1.3.2 Multiscaling (The Power Law Dynamics)

The distinguishing feature of the multifractal model is multiscaling of the return distribution's moments under time-rescaling. The Joseph and Noah effects can be observed because financial series follow a scaling distribution. A phenomenon satisfies the scale principle if a power law links all of the quantities relative to this phenomenon to one another. Power laws are interesting because of their scale invariance. That is, given a relation $f(x) = ax^k$, scaling the argument x by a constant factor c causes only a proportionate scaling of the function itself.

The property of invariance (up to a scale parameter) under aggregation of independent elements has been studied by Paul Levy (1925, 1937), and is called *L*-stability. In a linear process, the scaling function is fully determined by a single coefficient, its slope. It is thus called *uniscaling* or *unifractal*. *Multiscaling or Multifractal* processes allow more general concave scaling functions. In Fig. 1.5a and b, the price series examples from the SPX Index seem to obey the same law or appear to have been generated by the same process, whatever the chosen time scale. The movements in both graphs appear similar despite their different scales, showing that the price movements have the same structure whether we observe a week, month, year, or decade's period of trading.

Mandelbrot (Mandelbrot B., The Variation of Certain Speculative Prices, 1963) (Mandelbrot B., The Variation of Some Other Speculative Prices, 1967) followed by (Fama 1963) suggested that the shape of the distribution of returns should be the same when the time scale is changed.

Fig. 1.5 a Fractal nature of
the SPX index using a daily
scale. *Source*: Bloomberg.
b Fractal nature of the SPX
index using a weekly scale
for intraday sessions. *Source*
Bloomberg

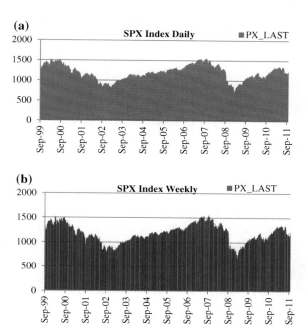

1.3.3 Pareto-Lévy Distributions

Mandelbrot discovered that financial series distributions are best represented by
Levy's laws, which are based on Pareto's findings. Pareto stated that the curve of
the distribution of revenue can be well represented by following the log normal
distribution, which differed from other distributions with their thicker tails. These
tails corresponded to the wealthiest 3 % of citizens and hid important information.
According to Pareto, the revenue distribution function is characterized by the
multiplier effect where the rich could more easily increase their wealth than their
poorer counterparts. The active feedback mechanisms underlying revenue distribution amplifies revenues, converting them into distributions with tails that were
thicker than normal, otherwise known as "fat tails". The Pareto-Levy law is a
generalization of the law of normal distribution, characterized by four parameters
and whose probability density function is represented as: Diagram 1.1

The volatility of any market is an instance of a "power law", that is, a correlation
in which the size of a price change varies with the power of the frequency of the
change. Empirical research shows that the actual power varies between 2 and 1 or
between "mild and wild" (Mandelbrot 2001, pp. 446–450). The properties of
market volatility are outlined as follows:

- When $\beta = 0$ and $\alpha = 2$. The curve has a power law of 2. In this case, we obtain
  the Laplace Gauss law of normal distribution:

$$\text{Log } (f(t)) = I. \delta. t - \gamma. |t|^{\alpha}(1 + I. \beta. (1/ |t|). \tan (\alpha. \pi / 2)$$

δ the **localisation** parameter of the average	γ the **scale** parameter which either compresses or dilates the distribution around δ	β the **asymmetry** parameter to the right (+1) or to the left (-1) $-1 < \beta < 1$ represents the skewness index, when β is nil, we get a symmetrical curve	α the characteristic **exponent** of the thickness of the tails $0 < \alpha\ 2$ the smaller the α, the thicker the tails.

Diagram 1.1 Pareto levy law (Mandelbrot 2008, p. 296)

$$\text{Log } (f(t)) = i .\mu. t - (\sigma 2/2). t^2$$

μ refers to the average and σ to the standard deviation. The variance exists only in this extreme case and is finite and stable.

- If $1 \leq \alpha < 2$. These distributions are stable for addition or invariant with respect to the addition of independent variables and to changes of scale, that is, the same laws[6] describe daily, weekly, and monthly price changes, for any time period (Fig. 1.4a and b). In volatile markets, variance is "infinite", with a "stable" average over time. As such, variance loses its importance as a risk measure. A value of $\alpha = 1$ corresponds to the Cauchy distribution, which has very fat tails.
- $\alpha \leq 1$. In this case, the average becomes unstable and can no longer be estimated.

1.3.4 The Hurst Exponent[7]: The Index of Long-Range Dependence

In the early twentieth century, the hydrologist Harold Edwin Hurst (1880–1978) worked on the Nile River Dam Project to determine the optimum dam size for the Nile

---

[6] It is worth noting here that Mandelbrot's inspiration started with his study of the distribution of cotton prices (Mandelbrot 1963): he noticed that instead of a bell curve, daily price changes yielded a different graph, with a large peak around zero and large deviations to the left and to the right, resulting from rarely occurring large price changes. Mandelbrot's own work on cotton prices and his other analyses of more recent data on equity markets, suggests that there are, in practice, far more significant price moves than the Neo-classical theory suggests. He established a mathematical framework which allowed him to model the incomprehensible variation of cotton prices that is far from the "polite Gaussian average" (Mandelbrot 2008, p. 169).

[7] The Hurst exponent (H) was named in honor of both Harold Edwin Hurst (1880–1978) and Ludwig Otto Holder (1859–1937) by Mandelbrot (2008, pp. 187, 297).

river volatile rain and drought conditions. In the 1950s, he noticed that the project was having problems not only due to the size of the floods but also due to their sequence: the Nile only floods a few times every 100 years but these floods occur in batches. A dam designed to prevent the average flood, as estimated by a normal distribution, would be highly unsafe and ineffective against the Nile floods as it does not account for extremes. Meanwhile, one cannot also build an infinitely tall dam. Hence, Hurst developed a method called the rescaled range analysis (R/S) in order to solve the Nile River Dam problem. The volume of a perfect reservoir (R) should determine how big the reservoir should be to avoid floods or droughts down the river; where

$$R/S = (a \times N)^H$$

- a is a constant.
- R is the difference between the peak and lowest values reached during a period of time; in finance, R would equal to the difference between the maximum and minimum prices.
- S is the standard deviation of the discharges from one year to the next;
- N is the number of years under study; and
- H is the power law exponent that drives the whole equation.

Hurst discovered that the charts for R/S relative to the empirical data were mainly made up of curves which were closely interwoven along a straight line. The slope of this line varied from one system to another. This was due to the structure of the correlation of the events in question. In fact, the R/S method measures how the distance covered by a particle increases as we look at increasingly longer time scales. A Brownian particle that travels one hundred seconds will get around ten times faster than the one that just travels one second—the distance travelled is proportional to some power of the time elapsed (Mandelbrot 2008). According to the laws underlying Brownian motion, the distance covered increases with the square root of time. As such, a series that increases at a different rate is not totally random; it exhibits a random walk with some bias. Hurst found out that in all natural cases the exponent[8] is above 0.5. In the case of the Nile River and other natural systems, H was close to 0.90.

Hurst's discoveries led Mandelbrot to develop a category of processes, wherein the processes are all self-similar or self-affine and have stationary and interdependent increments. The key exponent in the context of financial series is described as follows:

---

[8] Empirical research shows that although the exponent H(t) varies from one system to another, it remains stable for a given system, for whatever the time scale considered. To verify whether the Hurst exponent effectively measures the structure of the correlations of the series studied, choose any series with an exponent H > 0.75 and scramble its data. Then, calculate H again for the scrambled series and H will be equal to 0.5. This means that the system (memory) and its structure were completely destroyed by the scrambling operation.

$$\text{Log } \Delta P \ /\text{Log } \Delta t \cong H(t)$$

where P refers to price and t refers to time.

Mandelbrot concluded that H described the structure of the correlations of financial series and as such, becomes the characteristic exponent. In other words, H measures the long-term dependency of time series and their tendency to be cyclical, measuring, to a certain extent, the impact of information on the financial system. Because of the fractional values between 0 and 1 that the Hurst exponent (H) can take, Mandelbrot denoted the sum of this interdependent increment as a fractional Brownian motion (fBm). The value of H determines what kind of process the fBm is, that is, it gives an indication of the persistency degree affecting an asset price. A high H could indicate if persistent trend is at work, while a lower H may indicate a more random "classic" market force. The following outlines the scenarios for the different values of H:

- $H > 0.5$ indicates that the increments of the process are positively correlated and that the process exhibits long-range dependence and has a persistent fBm. Increases in price are correlated positively with each other (Joseph effect). If the trend is positive or negative, there is a strong chance that it will continue in the same direction. Mandelbrot (1997) explains this phenomenon:

  ... the persistence of the process becomes more and more marked when H increases from 0.5 to 1. In practice, this expresses the fact that cycles (non-periodical) of all kinds become more and more clearly distinct. In particular, slow cycles become more and more important (p. 29).

  When the expected value of the product of the increments is positive, the increments tend to be going in the same direction. This is the persistent fBm.

- $H = 0.5$ indicates that the process is in standard Brownian motion, which either obeys the Gaussian law or not. The observations are not correlated as past information has already been integrated in prices, as is often the case after a market crash.
- $H < 0.5$ indicates that the increments of the process are negatively correlated. It is an anti-persistent fBm. The series is characterized by its return to the mean and frequently shows abrupt reversals of tendencies.

The economic cycle is based on macroeconomic data as much as microeconomic data, as such different sector have different H exponents and economic cycles. For instance, the technology sector has a very high H exponent and a short cycle. This observation is understandable as technology companies must continually innovate in order to overcome competition and ensure their products do not become obsolete. The H exponent allows for the classification of the inherent risk of the time series. H is directly related to the fractal dimension (FD) which measures the roughness of a surface and is expressed as follows:

$$FD = 2 - H$$

One of the most compelling arguments against the Gaussian random walk is that markets appear to have fractal structures. The fractal dimension tells us what happens to the length, area, or volume of the fractal when you enlarge it. In general, a fractal that can be drawn on a surface has a value between one and two. For instance, a straight line or a smooth curve has a fractal dimension of one, while a full circle has a fractal dimension of two. The FD describes how a time series evolves; it is the product of all of the factors that affect the system that forms the series. As an example, the H exponent of an index is superior to that of its underlying stocks. Diversification can be optimized as the FD reads if there has been a reduction of the "noises" in the portfolio through calibrated allocations.

The notion of persistence explains the behavior of investors who analyze past prices using memory effects in the short- and medium-term and in doing so reinforces the memory effects in a reflexive relationship. That said, the persistence in the financial and economic systems is not and cannot be unlimited—bullish markets are followed by bearish markets and vice versa. This is illustrated by the convergence of H toward 0.5 at some point in time. It does not imply that the series is no longer fractal, but rather explains the instability of the system at that moment.

Although H can give an indication about a trend, or how much the price series is affected by memory, more research is still to be done to establish the reliability of the H values and validate their effectiveness in measuring efficiency and diversification. Mandelbrot (2008) echoes the same cautionary sentiment:

> We should be very careful and not rely on some ratios blindly (p. 263). My models would tell you about the tail exponent. It replaces the more familiar 'kurtosis' as an indication of how wild the price behavior of the index is. A score of 1.5 indicates that behavior is wilder than 1.6, etc. In addition, my analysis can give you an H-score, which is the exponent for long term price dependence—how much of a 'memory' the series seems to retain. And there are further intrinsic indicators that allow finer tuning (Wright 2005).

1.3.5 Multifractal Model of Asset Returns

In 1997, Mandelbrot, Fisher, and Calvet introduced the multifractal model of asset returns (MMAR) based on the research into multifractal measures by Mandelbrot (1972, 1974). MMAR incorporates important regularities observed in financial time series including long tails and long memory. The MMAR allows for long tails, correlated volatilities, and either unpredictability or long memory in returns, and thus combines the properties of many earlier models. Below is presented a concise explanation of MMAR (Diagram 1.2).[9]

---

[9] Multifractal measures were introduced in Mandelbrot (1972). A good understanding of multifractal measures can be found in (Mandelbrot, Fisher, and Calvet, A Multifractal Model of Asset Returns, 1997) paper. This paper introduces the concept of Multifractality to economics as it focus on a very concrete aspect of Multifractality.

Multifractal Model of Asset Returns (MMAR)

•Long-tails
BUT it does not necessarily imply an infinite variance of
returns over discrete sampling intervals

•Long-dependence
which is a characteristic feature of fractional Brownian
motion . BUT it incorporates long memory in the absolute
value of price increments, while price increments themselves
can be uncorrelated (allows the possibility that returns
themselves are white) in the absolute value of returns and
that volatility can have a Long memory characteristic.

•Scale consistency (time-invariance)
meaning that a defined scaling rule relates returns over
different sampling intervals. the price of a financial asset is
viewed as a Multiscaling process (nonlinearity of the scaling
function)
•Trading time
a random distortion of clock time that accounts for changes
in volatility.

Diagram 1.2   Multifractal model of asset returns (MMAR).

* fBm was introduced by Mandelbrot and van Ness (Mandelbrot and Ness,
Fractional Brownian Motion, Fractional Noises and Application, 1968).

** White noise: There are no linear dependencies in the conditional mean.
Compatible with the martingale property of returns (an unbiased random walk is
an example of a martingale).

***Trading time introduced by Mandelbrot and Taylor (1967) (Mandelbrot and
Taylor, On the Distribution of Stock Price Differences, 1967), is the key concept
facilitating the application of multifractals to financial markets. The MMAR posits
a trading time that is the cumulative distribution function (c.d.f.). of a multifractal
measure. Thus, trading time will be both highly variable and contain long memory.
Both of these characteristics will be passed on to the price process through
compounding. The main significance of compounding is that it allows direct
modeling of a processes' variability without affecting the direction of increments
or their correlations. Properties 3 and 4 are consequences of the definition of
multifractal processes.

****In fact, the MMAR is not the first to combine long memory with a mar-
tingale property. The FIGARCH model of Baillie, Bollerslev and Mikkelsen
(1996) (Baillie Bollerslev and Mikkelsen 1996) also has this feature. The most

important distinguishing feature of the multifractal model is that returns are scale-consistent, while FIGARCH is not. Like FIGARCH, the MMAR incorporates long memory in volatility. In addition, the MMAR allows the possibility that returns are uncorrelated, but does not require it. This is an important property for researchers interested in issues of market efficiency. The main advantage of the MMAR over alternatives like FIGARCH is the property of scale-consistency. Because of this property, aggregation characteristics of the data (otherwise thought of as the information contained at different sampling frequencies) can be used to test and identify the model.

One the main features of financial markets is the alternation of periods of large price changes with periods of smaller changes. Fluctuations in volatility are unrelated to the predictability of future returns. This statement implies that there is autocorrelation structures dependence in the absolute values of returns. The multifractal model of asset returns combines the properties of L-stable processes (stationary and independent stable increments) and fractional Brownian Motions (tendency of price changes to be followed by changes in the same (or opposite) direction) to allow for long tails, correlated volatilities, and either unpredictability or long memory in returns.

In summary, information is neither totally nor immediately integrated in market prices, but manifests itself as a bias. This bias continues until the arrival of new information that changes the bias' magnitude, direction, or both after a given number of observations in a series. This explains the convergence of H toward 0.5, which suggests that the system follows another process after a certain time T.[10] T is a useful but unpredictable parameter in economic and financial analysis:

> Fractal modeling falls apart only when the units measured are less than two minutes or longer than 180 days, a breakdown that Mandelbrot compares to the collapse of the normal laws of physics at the tiny quantum or the gigantic cosmic level (Thompson 2004).

In Chapter 2 an additional process is introduced to try to explain what happens at time T.

---

[10] In the case of the solar system, it is the time necessary so that an error in positioning or a disturbance of motion can be multiplied by 10. Thus for a period where: $t < T$ we can follow the system by the calculation and $t > T$ we completely lose the trajectory of the system.

Chapter 2
The "Noisy Chaos" Hypothesis

Abstract The concept of "noisy chaos" is introduced in this chapter, based on the processes of bifurcation, entropy, and convergence which occur at the heart of the instability in the financial markets and take into account the sensitivities of the financial systems. The key here is to identify and measure indicators that allow us to construct a model that accounts for extreme consensus factors (undisseminated information, correlated investment horizons, and high leverage) in estimating market reversals. Instability is a relatively subjective notion. If we think in calendar time, the system is unstable as the daily fluctuations seem erratic when compared to periods of months and years. But if we think in intrinsic time, it is as if we are looking at the week as a year, the day as a month, and the minutes as days... in doing so, and as markets are "self-affine" then the L-stable process can be found at the day level and the erratic fluctuations will be at the seconds level.

> Overshoot then crash, Joseph effect, then Noah effect, again and again. How big is the overshooting? It can only be estimated after the fact... but that is little consolation to anyone living in the real markets... we can never forecast when the bubble will burst: result prices gyrate, from boom to bust, from bust to boom (Mandelbrot 2008, p. 204).

2.1 Financial Systems at the Edge of Chaos

The discovery of chaos originated from the observation that some dynamic systems can attain autonomy over the course of their evolution. In 1961, the meteorologist Edward Lorenz discovered that this evolution is very sensitive to initial conditions.[i] The meteorological system is regulated by a chaotic dynamic, characterized by the existence of an amplifying effect, also called the "butterfly effect". Lorenz (1963) states the following:

Y. Hayek Kobeissi, *Multifractal Financial Markets*, SpringerBriefs in Finance,
DOI: 10.1007/978-1-4614-4490-9_2, © The Author(s) 2013

Can the flap of a butterfly's wing in Brazil set off a tornado in Texas? The sensitivity to initial conditions finds its source in the non-linear character of systems, where the effect can be more proportional than the cause: these considerations explain why the validity of weather forecasts stays limited to several days.

Lorenz's weather forecasting experiments demonstrated that after a certain period of time, all nonlinear dynamic systems that have attained a chaotic regime once again pass through neighboring states of paths they have already gone through. The trajectory of these systems has a tendency to occupy only a limited part of space because of its stationary characteristic and to converge toward a "strange attractor" of a fractal nature.[1] We cannot pinpoint their exact location on the strange attractor; it can only be determined *a posteriori*.

The attractor is defined as the level of equilibrium of the whole system. If a system is headed toward stable equilibrium, its equation is linear and has only a single solution called the "point attractor". When the system is regulated by a nonlinear dynamic, the system's trajectories are considered orbits in its probability space. The various points through which an orbit passes are the solutions of the equation. The solutions can be calculated and forecasted because the system always follows the same periodic orbits. In this case, the attractor is called the "limit cycle." For some systems, however, the orbits are different each time, that is, the trajectories never pass through the same points. The orbits are nonperiodic and remain confined in their probability space; the attractor is called the "strange attractor." The existence of different attractors demonstrates the complex mathematic phenomenon of deterministic chaos, that is, systems that appear to be totally random yet governed by a deterministic process. Ekeland (1995) wrote:

> Chaos theory considerably enlarges the possibility of using deterministic models... To propose a deterministic model is to leave a space for randomness, a dimension of the unpredictable... The system is confined to its strange attractor, certainly, but its movement around the attractor escapes us, which is a prediction that we can make beyond the characteristic time... an admirable and subtle dosage of randomness and necessity (p. 101).

In essence, chaos appears in relation to self-reinforced mechanisms generated by positive feedback loops. Each of these loops can contribute to the auto enforcement of a phenomenon until it is curbed by other mechanisms. When the parameters of a system are stimulated beyond a threshold limit, an important and violent change in the dominance of the different positive loops is produced, a phenomenon known as "bifurcation".[2] From its initial equilibrium state, a system bifurcates in a very orderly fashion until it reaches a chaotic state. According to Mandelbrot, strange attractors make it possible to consider both the Joseph and

---

[1] Mandelbrot defines fractals as strange attractors of all deterministic chaotic systems. These objects possess the particularity of having an infinite perimeter while being contained in a finite space. Mandelbrot does not assert that markets are chaotic but that the outcome of any process is sensitive to its starting point (Mandelbrot 2008).

[2] The Australian biologist, Robert May, was one of the first to discover bifurcations in 1976.

Noah effects. The fractality of the system, due to its self-affine characteristic, allows cycles to be identified; these limited nonperiodic cycles depend on the process at play and on the behavior of the market players. Historical analyses have shown that economic cycles cannot be predicted; that their ends cannot be determined and that they are not random. There are perfectly identifiable logical linkages between market players, their expectations and behavior, such as, for example, investors and the US Federal Reserve, their actions and the consequences of their actions on the interest rates.[3] The cycles are a consequence of the complex system that regulates the economy.

The L-stable process depends on the structure of the correlations between prices. The characteristic time, named hereafter the bifurcation moment (B), however, poses a strict limit for all quantitative forecasting, such that beyond it, the system acquires a certain degree of freedom.[4] The evolution of a financial series at characteristic time B is completely erratic; as time passes. No trend can be clearly defined as H gets closer to 0.5; it is as if the memory of the system has faded. Although, the financial process seems to follow random motion, the system still retains its fractal nature and so stays confined to its strange attractor.

2.2 Constrained and Optimal Entropy Model in Noisy Information Systems[5]

The concept of constrained entropy (Dinkel 1979) refers to the fact that at all times and at different degrees there is noise constraining the speed of information dissemination in the financial markets systems and that information arrives following a sporadic jump process (Black 1976). The speed of the price adjustment process is constrained to some finite speed by taxes, transaction costs, information costs, assimilation time (the costs of acquiring and analyzing information[ii]), and divergence in perceptions. Finite speed of information dissemination and assimilation in the marketplace and discontinuous information process lead to disequilibrium conditions and prevent entropy from being maximized. Equilibrium and disequilibrium conditions in such a market are derived from Nicolis and Prigogine[iii] (Nicolis 1977): if we consider a variable (gt) as the amount of (undisseminated) information in the market, it acts as a measure of disequilibrium, i.e., the greater the amount of undisseminated information, the greater the divergence from

---

[3] This concept is discussed in details in Chap. 4.

[4] Time is characteristic in the sense that it depends on the system in question and varies according to the stock, sector, and microstructure of the market. If the characteristic time of the solar system is 100 million years, then the forecasting time would be 1 month.

[5] Ideas gathered from Pr. D Nawrocki research papers to fine-tune and support the concept of Noisy chaos proposed in this book (Nawrocki 1984).

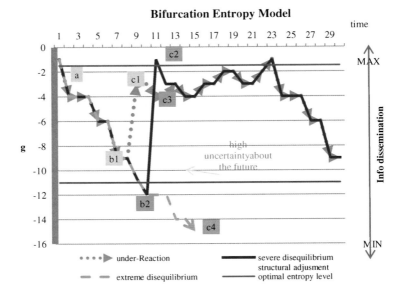

Fig. 2.1 Bifurcation entropy model

equilibrium. The bifurcation entropy (BE)[6] model under different scenarios is explained in Fig. 2.1 and in the following subsections.

2.2.1 Instability at n [Milliseconds; Days] and High Frequency Trading

To verify the stability of the distributions of price changes, Peters (1996) compared the H exponent of daily distributions for a daily series from the S&P 500 index. While carrying out the stability tests, Peters observed that the H values increased progressively from 0.59 to 0.78 between $t = 0$ and $t = 30$ days, after that it fluctuated between 0.78 and 0.81. Peters attributed the lower values of the H exponent to the existence of noise in the system. This noise disappeared after

---

[6] BE can be thought of the verb "be" referring to the existence/survival of the system.

approximately 30 days, which indicates that short-term price fluctuations are characterized by a very short memory, otherwise known as the Markovian memory, which is an autoregressive process characterized by a very short-term dependency.[7] This implies that for n [milliseconds; days], there is an exponent H_ε [0.5; H]. During this period the series has more noise which disappears after a very short time. As such, this type of dependency must not be confused with long-term dependencies that stay in the system and where present events affect the future. To illustrate a short-term dependency where erratic daily movements have no effect on future prices, we take as an example the sudden stock price increase when the company's CEO is being interviewed on TV. During this period of time (approximately 10 min) the price can increase by 5 % or more. Afterwards, the price returns to its initial level or its value prior to the interview. This explains why volatility is higher in the short-term than in the long-term.

A more general cause of erratic daily movements is the high-frequency trading (HFT) effect. An intraday is viewed as a cycle itself, where the investment horizon can vary between milliseconds and hours. On May 6 2010, Wall Street experienced a temporary blackout. The Dow Jones Industrial Average plunged around 700 points in minutes before recovering. Everyone suspected that HFT aggravated this sudden decline, although no one confirmed it. HFT strategies can vary from providing liquidity (rebate trading and market making), trading the tape (filter trading and momentum trading), and statistical trading (statistical arbitrage and technical trading). Critics say that HFT's powerful algorithms equipped with ultra-low latency solutions give high-frequency traders an unfair advantage by detecting on time markets imbalances within nanoseconds and then executing millions of orders in less than one millisecond,[8] HFT takes advantage of price movements, leaving other investors behind the curve. That said, most of these types of trading is legitimate, reducing trading costs for ordinary investors and ensuring market liquidity. Problems occur only when strategies are used to confound other investors by issuing then cancelling orders almost simultaneously. In this case, regulators should monitor trading activity and penalize such trades, for example, when The Financial Services Authority (FSA) successfully pursued online brokerage company Swift Trade for manipulative trading layering[9] that triggered a series of small price movements on the London Stock Exchange between January 2007 and January 2008 and from which Swift Trade profited about £1.75 million (Library 2011).

We would like to point here that instability is subjective. If we think in calendar time, the system is unstable as the daily fluctuations seem erratic when compared to months and years periods. But if we think in intrinsic time, it is as if we are

---

[7] The function can have the following form: $X_{(t)} = X_{(t-1)} + \varepsilon$, with ε a random Gaussian variable with null average.

[8] Sub-microsecond (or nanosecond) delivery for level 1 markets and end-to-end through put of only 2 microseconds for complex markets. NovaSparks.

[9] In trade layering, orders are entered onto an exchange to artificially improve the price of a stock. Once this is achieved, the orders are then cancelled.

looking at the week as a year, the day as a month, and the minutes as days... in doing so, and as markets are self-affine[10] then the L-stable process can be found at the day level and the erratic fluctuations will be at the seconds level.

2.2.2 Lévy-Stable Process

At $n < T$ (n is the time and T is the characteristic time), the evolution of the markets follows a fractional Brownian motion (fBm). The Brownian motion is biased, the probabilities of distribution of the price changes lean toward one direction. Prices are generated by a persistent, fractal, Lévy-stable process. This process is determined in part by the structure of the correlations of the series measured by the Hurst exponent and the positive feedback or reflexive relations synchronized by the attitudes of the financial players who rely on technical analysis. In addition to the L-stable process, another process is at play, one that is regulated by a Markovian memory. However, the L-stable process dominates the Markovian one. Each individual has his own perception and makes his decision according to that perception. These heterogeneous perceptions create some uncertainty and differences, albeit minimal, between the market price and the fundamental price.

In the short- and medium-term, the financial markets are dominated by investors who have appropriate investment horizons characterized by their own processes, the integration of which leads to Lévy distributions. As we go further in time, this process loses its pertinence. In other words, when the system's memory is not backed up by the arrival of confirming information or on the contrary, by divergent information, its effect decreases until it becomes impossible to measure. Sequentially this is what happens referring to Fig. 2.1.

1. a: Arrival of new information
2. Lags in information process creates greater disparity between the equilibrium price and the actual price
3. Arbitrageurs enter the market in greater volume as they can generate enough profit to cover transaction and information costs
4. Higher volume of trading
5. a-b1-c1: gradual increase in the speed of info dissemination, reinforcing the market trend. Gradual but not continuous (stable) assimilation of information over time (memory effect). Autocorrelation varies with the amount of disseminated info and its speed. The speed of information dissemination increases so that undisseminated information does not have a chance to accumulate to a bifurcation point. Within the context of the bifurcation model of conditional market returns proposed by David Nawrocki and Tonis Vaga (David Nawrocki 2011), in this case returns execute a biased random walk and

---

[10] Re. Chap. 1.

if the mean return exceeds the associated costs, the markets may offer profitable trading opportunities.

2.2.3 Bifurcation Entropy Process

When undisseminated information accumulates to a Bifurcation point, it leads to increased uncertainty and widening of the competence-difficulty (C-D) phenomenon.[11]

Following a shock, the system moves to an unstable state where the margin of trading and the trend are broken up. Shocks are produced at the positive feedback loops of the system, when a parameter is stimulated beyond a certain critical threshold. Laszlo (1987) defined a theoretical framework for the analysis of markets, categorizing the parameters that influence the dominance of economic system loops into: technological innovations; conflicts and conquests; and economic and social disturbances such as shortages and financial crises.

For instance, after the introduction of technological innovations, a shock is produced as scientific discoveries stimulate productivity and profit margins of leading technology companies. Multiplier effects, self-reinforcement of growth expectations, and other factors also stimulate growth throughout the economy. During this period of growth, however, key financial players, such as investors, analysts, and regulatory authorities will not have the appropriate valuation tools to analyze and assess the risk of new financial and economical changes. The difference between the stock market price and its fundamental value at the beginning of the period can be justified by the economic growth resulting from the innovations. It is human nature, however, to overestimate the potential growth and advantages that innovations will bring to the economy, which leads to an overvaluation in the equity market (overconfidence and gain seeking taking over loss aversion is human nature). On top of that, social contamination encourages the progressive formation of a highly positive consensus in the market. The overvaluation is amplified when profits get reinvested in the markets. The overvaluation bias continues to grow until it reaches a "consensus extreme" and becomes unsustainable. Shocks are amplified by the actions of the economic agents themselves. As financial conditions worsen, market participants risk aversion increases. The increasing difference between reality and perception is adjusted through the phenomenon of bifurcation by a

---

[11] This is the spread between the competence of the investors and the complexity of the information. (Nawrocki 1995) (Heiner 1983) (Kaen and Roseman 1986).

radical change in perception through which financial disengagement begins. It is then necessary to add consensus extreme as a fourth parameter to Laszlo's list.

A high consensus level is a tangible sign for the beginning of a major market reversal. Let's take an example of two states of price equilibrium. State 1 is where we have equal notional amounts on each side of the market (buying/selling) and where buyers and sellers are of equal size and quality. State 2 is where we have equal notional amounts on each side of the market (buying/selling) and where buyers and sellers are of different size and quality. The difference between state 2 and 1 lies in the quality and size of the buyers/sellers. Some practitioners suggest a contrarian[12] approach wherein one acts contrary to the market consensus peak. According to Tvede (2000, p. 117), if the majority of the market is buyers, the average seller must be larger than the average buyer in order to fulfill the large demand. If 90 % of the market is buyers, then sellers must be about 10 times larger than the average buyer; if 95 % of the market are buyers, then the average seller must be 20 times larger than the average buyer. As such, it is important to know who is behind the trades. What would you think if you know that the current market price is heavily sold by few big entities while large number of retail investors is still buying it? As long as the big players can find counterparts to their trades they will execute them and decrease their long exposures. When the numbers of small traders decrease, traded volume decreases exponentially to smaller and smaller levels. Prices start going south to get trades executed, small players end up aligning willingly (perceiving finally the risks ahead) or unwillingly (due to margin calls) with sellers' original view.

Investment horizons alignment on the short side is the prelude to a crash, while alignment on the long side is the prelude to a long rally. Either way it leads to tail risks. To clarify the idea of consensus we outline here that is not only in the market direction but in the investment horizon. Would it be possible to know in advance when this alignment is becoming dangerous? To do so, it would be wise to start by understanding the dynamic behind investment horizon alignment; that can be:

- The emergence of short-term flash traders, such as hedge funds; mixed signals from clients; short-term incentive systems; HFT.
- Behavioral biases, such as herding and recency biases.
- Shorter period of management mandates (3 years).
- Volatile markets and changing macroeconomic conditions.
- High turnover and volatility cycle: this is a cycle as we can wonder if the market volatility pushes higher turnover, or does higher turnover lead to increased market volatility?

---

[12] It is important to note, however, that the success of the contrarian approach depends on timing, which can only be determined through guesswork. The adoption of Bayesian principles might help as well, where the probability of the occurrence of an event is related to the occurrence or nonoccurrence of an associated event. In this case, it would be wiser to base one's decisions on the analysis of the roughness and the level of risk in the market.

It becomes obvious that the main lead to follow, in order to identify signs of alignments, is the shift in market liquidity i.e. the flow of money between sectors, industries, strategies, investment products, and the portfolio turnover level in some style type strategies. This is a clear explanation of how and why correlations tend to 1 at panic juncture.

Now to integrate the above-mentioned phenomenon we proceed as follows:

As stated in the fractal market hypothesis, the distribution of price changes is marked by a large number of small fluctuations.[13] An important question arises concerning a third process which produces these fluctuations.[14] The passage from an L-stable state to an unstable state is the result of a long process set free parallel to the L-stable process. An understanding of this process requires the review of the entropy notion.

At any point in time, a system is distributed across N microstates, each denoted by i, and having a probability of occupation, p_i, and energy, E_i. These microstates form a discrete set. The internal energy is the average of the system's energy, expressed below:

$$U = \{E\} = \sum_{i=1}^{N} p_i \, E_i$$

This is the microscopic statement of the first law of thermodynamics. A macrostate characterizes observable average quantities, such as temperature, pressure, and volume. When applied to finance, the macro-factors are indicators, such as volatility indices, unemployment, and inflation. The microstates define a system's molecular details including the position and velocity of each molecule. When applied to finance, it defines financial agents' behavior, including reaction time, intensity, and the tendency for under- or over-reaction. Entropy measures the degree to which the probability of the system is spread out over different possible microstates. It is expressed by the second law of thermodynamics as follows:

$$S = -K_B + \sum_{i} p_i \, \ln p_i$$

where k_B is the Boltzmann's constant.[15]

---

[13] This large number of small fluctuations is different from those at the base of the Markovian process, of which the effects, while they are of short duration, have been totally integrated into the system.

[14] Under-reaction leads to over-reaction at time T owing to the information ignored due to the prospect and disposition effect. The tendency of some investors to hold on to their losing stocks creates a spread between a stock's fundamental value and its market price, as well as price under-reaction to information.

[15] Boltzmann's constant defines the relation between absolute temperature and the kinetic energy contained in each molecule of an ideal gas. It is named after the Austrian physicist Ludwig Boltzmann (1844–1906), and is equal to the ratio of the gas constant to the Avogadro constant: $K = R/N_A = 1.3807 \times 10^{-23}$ J.K^{-1} (joule per Kelvin).

Based on the various definitions of entropy on a macroscopic scale,[16] and in fields, such as statistical mechanics,[17] cosmology,[18] and information theory,[19] we can postulate that: Entropy is an indicator of the level of undisseminated information in a system after taking into account observable macroproperties. Not all information is readily integrated into the market price. The lost/ignored or undisseminated information in the L-stable process is therefore assimilated into another endogenous process, the bifurcation entropy process (BE process). This process evolves until the arrival of supporting information (the amount of "lost" information) that hits a threshold which triggers a shock in the financial system and takes over the L-stable process. The shock depends in part on the relative importance of the arrival of new information as well as on the accumulated information previously. A glass of water overflows not only because of the last drop of water, but because of the drops of water that have accumulated. Overconfidence feeds the BE process. When too much information is ignored (because financial agent react only to comforting news), the market gets into a high consensus state. Severe market disequilibrium (Boulding 1981) (Thom 1972.) (Nicolis 1977) are observed leading to the following extreme scenarios:

7. Market over-reaction: unable to disseminate the valuable info, the financial system triggers the bifurcation phenomenon (g hits a bifurcation point b2). A mean regressive force propels the returns back toward the mean after over-reaction to prior news (Nawrocki and Vaga 2012). The market restructures itself to improve the flow of information and the speed of dissemination to reach a stable comfortable disequilibrium condition a-b2-c3 that can ensure the system survival. In the case of the nonlinear dynamic financial system, this process obeys a power law and is exponential, so very abrupt.
8. Special not viable case c4: no or few information is disseminated; the markets fail to properly allocate financial resources. The disequilibrium becomes so severe that investors and borrowers refuse to participate in the market. This action by the participants threatens the survival of the market system.

At bifurcation time, overwhelming disequilibrium triggers large fluctuations as uncertainty about what has been missed is shaking the markets' participants. Unable to assess the fundamentals, their view is blurred with no clearness about the trend ahead. In short, domino effects are observed where there are losses and

---

[16] On a macroscopic scale, entropy is a function of thermodynamic variables (such as temperature, pressure, or composition). It is a measure of the energy that is not available for work during a thermodynamic process. A closed system evolves toward a state of maximum entropy.

[17] In statistical mechanics, entropy is a measure of the randomness of the microscopic constituents of a thermodynamic system.

[18] In cosmology, entropy is a hypothetical tendency for the universe to attain a state of maximum homogeneity in which all matter is at a uniform temperature, or a state known as "heat death".

[19] In data transmission and information theory, entropy is a measure of the loss of information in a transmitted signal or message.

bankruptcies. Consequently, uncertainty about the future reaches its maximum, volatility reaches new highs and correlations tend to 1.

Complexity occurs at the edge of chaos to invoke self-organization,[20] thus defining new structures insuring the survival of the financial system. In this process, the economic system evolves from a complex state to an even more complex one, after the disengagement, oscillating between equilibriums, the number of which increases exponentially. This phenomenon manifests itself through bifurcation; the market consensus is divided as uncertainty broadens. The market oscillates between several equilibriums, driven by uncertain perception, which is the predominant variable here. The entropy process takes into account all unstable subjective variables. At $t = B$ bifurcation point, the crucial indicators are the fundamental financial valuations which depend on economic parameters (the strange attractor) which are uncertain at this moment.[21]

9. Following the bifurcation, the market over-reacts and the entropy process takes over hastily. This is observed in Fig. 2.1 at point c2. It is a special point referring to Optimal Entropy. Full dissemination of information refers to maximum entropy, and occurs when the speed of information dissemination is infinite (Cozzolino 1973). Optimal entropy refers to optimal information dissemination limited to transactions and tax costs. It is the maximum entropy relaxed under finite speed conditions where no enough arbitrage opportunities are available and the state is close to a random walk. Once bifurcation is triggered in the system, the c2 point hit the optimal entropy level line that is the optimal equilibrium state following the abrupt (in intensity and hasty reaction time) increase in the speed of information dissemination. The market is then said to be effective. According to Nawrocki and Vaga working paper (David Nawrocki 2011).

...An efficient market occurs at the equilibrium points Rbull and Rbear. At these points the average conditional return is equal to zero and on a short-term basis, conditional returns fluctuate in a random walk around zero due to the new information arrival process. This is where there is no undisseminated information. The market neither over-reacts, nor under-reacts. It is neither mean regressive nor trend persistent at these dynamic equilibrium points, where conditional returns operate as a random walk in response to new information arrival (zero mean)...

Entropy depends on the probabilities of the microstates—the more microstates are available to the system, the greater the entropy. Opinions are divided about what should be the market fundamental value. Uncertainty is conditioned by the

---

[20] According to M. Barranger "Emerging behavior results from global interaction between the scale's constituents, the combination of structure and emergence leads to self-organization" (Barranger 2000).

[21] To determine the trajectory of a deterministic system, it is enough to know its initial position. For two different initial positions, there are two different trajectories. A system is chaotic if it amplifies, even a little, the initial difference, and attains freedom, however, limited by its attractor. It is in this amplification of short durations that randomness is found.

ability to make future prediction. At this stage, the market follows a less biased random walk and uncertainty about the future trend poses a limit. The Hurst exponent moves closer to 0.5.

These differences are then amplified by subjective biased investor's perceptions. The increase in the microstate possibilities means more randomness; the more efficiently the information is integrated into the market prices, the smaller the difference between fundamental value and market price, but also means high uncertainty about the future trend of the financial market. This uncertainty is inherent to the interpretative or perception models. Note that the characteristic time B is unstable because it depends on the moment where the critical consensus level is reached and on the system's initial conditions. This is like the critical point in phase transition cited by Barranger (Barranger 2000), the place where long range correlations are most important. As with correlations, T varies according to circumstances and is a function of the attitudes of the market players and their perception of risk; as new categories of investors appear with each cycle; the degrees of aversion to risk also change. Accordingly, it is difficult to establish preset relations which would allow the forecasting of the occurrence of financial disengagement.[22] Investors do not react in the same way; and when they do, an extreme consensus is reached. On top of that, we cannot accurately predict the reactions of individuals especially during periods of panic. The following are some of the indicators for the consensus level estimation:

- Volume strength analysis and market liquidity: an indication of the flow of money (energy) in or out of the market open system. Volume can validate[23] a price, infer liquidity, give weight to new information, and reveal convictions and market disparity.[24] In brief, if price changes reflect the average change in market expectations, trading volume reflects idiosyncratic reactions across all traders (Verrecchia 1991).
- Correlated/diversified investments horizons (liquidity shifts and portfolio turnover levels).
- Correlation, volatility, and leverage level[25]: The volatility index or the Fear index as its spike translates the Noah effect born from fear, market withdrawal, and consequent illiquidity. When volatility and correlation levels are too low; it is an indicator that the market is overleveraged and most of the time ignoring contrarian information. This is usually followed by a sudden rise to historical

---

[22] Management techniques which are based on the analysis of fixed ratios, like the PER (Price earnings ratio), are often found to be too premature.

[23] Here, it is important to look out for the misrepresented current volume flows due to high-frequency trading (HFT) strategies such as tape trading (filter trading and momentum trading) and statistical trading (statistical arbitrage and technical trading). Volumes emanating from these strategies have to be deducted.

[24] For an in-depth analysis on volume, read Buff Dormeier Investing *with Volume Analysis* (Dormeier 2011).

[25] Ideas backed and found in (Bookstaber 2011), (Nawrocki and Vaga 2012), (Nawrocki 1984).

levels. Following bifurcation the market will be characterized by a mean jump process as well as non-stationary covariance process.

- Put to call ratio (open interest): This ratio is directly derived from options trading and is used by options traders around the world as a contrarian indicator of market sentiment. But it lacks over the counter (OTC) options.
- Three month volatility skew: for instance, it illustrates the fact that the implied volatility is higher as put options go deeper out of the money (strikes below current market levels). Volatility skew may reflect investors' fear of market crashes, which would cause them to bid up the prices of out of the money options.
- Volatility term structure: This indicates how the skew will respond to the passage of time and swings in the underlying values.
- Hedge funds net stance: This captures the difference between the proportion of hedge fund holdings in long and short positions. On 5 August, 2011, this stance was at 40 % long instead of a high of 70 %; during bear markets, it can fall to as low as the mid-teens.
- Tracking of funds and hedge funds leverage level to trace their level of engagement or disengagement.
- Borrowed shares: This is an indication of short selling. According to data compiled for Bloomberg by the London-based research firm, Data Explorers, borrowed shares have risen in August 2011 to 11.3 % of stock available for lending, from 9.5 % in January during the same year.
- Credit Default Swaps (CDS): When the price of a credit default swap goes up, it indicates investors' perceptions that default risks have risen.
- Bid-Ask Spreads of prices. Specially those of the debt market.
- Bond yields: Falling bond yields is an indicator that the economy is slowing down and trouble is ahead for stocks markets.
- Treasury market …

Figure 2.2 summarizes the noisy chaos process.

*High volatility seems to destroy trend persistence. It is the critical temperature above which the dynamic disequilibrium ordered state vanishes and give way to a mean regressive state. **Indicators of correlated investment horizons consensus.[26]

Portfolio turnover
Some management style rely less on momentum and shorter term drivers than other style types might. A level of portfolio turnover higher than expected indicates a shift to short-termism.

---

[26] As per Dr. Brock book (Brock 2012):

"it turns out that what matters to market prices and volatility is not merely that people's bets are wrong, but additionally how *correlated* their mistakes are. It is when most investors are wrong in the same direction that the greatest portfolio adjustments occur, and that prices change the most; for most everyone will either sell or buy, and the price swings accordingly… A classical example of a "correlated mistake" where bets on US house prices as late as 2006. Almost no one assigned high probability to the inconceivable 35 % drop in house prices that transpired. The market response was fast and furious".

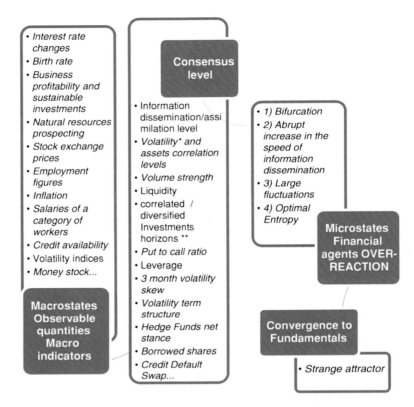

Fig. 2.2 Noisy Chaos (Consensus indicators align in the same outlook → maximum consensus is reached. Microstates start bifurcating into multiple opinions that is different probability spaces. This triggers collisions in market velocity, where reactions intensify as players ruminate their trade decisions; implying higher entropy and so higher uncertainty is observed)

- value managers
- large cap strategies
- (SRI) strategies

Liquidity shift

- Style type
- Equity life cycle (starts-up, star, cash cow, dog)
- Fixed income duration
- Hedge funds strategies flow (event vs. relative value)
- Private equities
- Mergers & acquisitions
- Safe heaven investments

Understanding market events and their effects on the market system is an essential element toward building a financial warning system. The key here is to identify all pre-switching points (amplitudes magnitudes = max consensus or before, max entropy is too late); implement it in algos that can trigger a warning signal beforehand.

2.2.4 The System Learns, Changes, and Converges

The chaos theory refers to the identification of some order in the nonlinear dynamic process regulated by uncertainty. Order is observed when the system, although in a state of freedom, remains confined to its strange attractor. Freedom and uncertainty remain limited to the fractal nature of the series. Converse, a series that follows a random walk has no particular shape; it evolves in the same way gas takes the shape of the container in which it is placed. These random series do not hold any correlation structure that would permit them to evolve while retaining their dimension, contrary to economic and financial series that do at different levels. Fractals retain their dimension in whatever space they evolve; In *Le Chaos*, Ekeland (1995), notes:

> Admirable and subtle dosage of randomness and necessity! Resolved here in one blow is a whole army of false problems concerning human freedom in a deterministic universe. We don't see, as Laplace did, a free sky opening onto an infinite horizon, so clear that we have the impression that we can touch it. Nor is it an open sky, full of fog that catches our gaze and reveals to us the entire horizon. What we see are both together, like a sky full of rain, where the gusts of wind give us glimpses toward the horizons far away where the sun is shining. (p. 104)

As we are not in a closed system, entropy evolves from low to high levels and vice versa according to financial agents' perceptions. After the market collapse, investors' reaction to news and information is almost simultaneously integrated in the stock market. The main participants are the surviving professional investor, the gap between the fundamental value and the market price diminishes until the arrival of enough information to trigger again a new cycle. The market converges to its attractor, the macro fundamentals. Some period of time must elapse before we can identify the new tendency of the market and evaluate the new values for the L-stable process variables.

> ...A bifurcation parameter, which is related to the slope of the conditional return map in the neighborhood of moderate prior day returns, illustrates that the bifurcation dynamic may change over time. Therefore markets can be viewed as restructuring their information structure both on a short-term basis in response to volume and volatility and on a longer time scale as a function of changing market conditions such as the advent of negotiated commissions and computerized trading. (Nawrocki and Vaga 2012)

Beyond its characteristic time B, the financial system cannot be predicted with certainty but it still retains its fractal nature. The system attains a certain degree of freedom, a kind of complex system prevails after optimal bifurcation entropy (BE) has been reached, and during which it moves back toward its strange attractor.

Fractal theory allows us to uncover the dynamic that regulates the financial system which in turn allows us to identify the underlying processes including the L-stable process, entropy, and convergence. The concept of noisy chaos considers traditional approaches as cases of a more general dynamic approach. Today's managers need to think beyond the classical paradigm. Their aim should be to achieve uniqueness and innovation, to survive in ever-changing financial markets, and to deliver performance through market cycles. As Oscar Wilde said:

Be yourself, everyone else is already taken

Subjective interpretations lead to economic and financial unpredictability. In the next chapter, we focus on the human behavior a critical variable of the financial process. The aim of chapter three is to be aware of the diversity of interpretations after the arrival of information in the market. Behavioral finance tries to group these different behaviors but that doesn't mean that we can systematically and at all times be certain of market behavioral interpretation that we make. Each one of us ends up guessing according to his perception. Being aware of that allows us to build a shield layer against ourselves and consequently build a strategy while always estimating a margin of error.

Chapter 3
The Mind Process

Abstract The sense of time and financial players' behavior is the central theme of this chapter. The notion of "intrinsic time", a dimensionless time scale that counts the number of trading opportunities that occur regardless of the calendar time that passes between them, is explained to highlight the difference in investors' perceptions and how to use this fact as a tool in understanding the processes at play and the biases to identify and avoid. As new information is constantly entering the market financial participants need to revise their expectations according to their own utility perception. As such the study of utility is important to understand the financial marketplace. The key element in any information content is the surprise element. Surprise is experienced only if an unexpected outcome occurs from which a new or different utility per individual is derived. Bearing in mind that information is a decreasing function of probability, we introduce an innovative subjective utility theory as per the findings of Viole and Nawrocki: the "Multiple Heterogeneous Benchmark Utility Functions". Bayes' theorem and fuzzy logic that has found application in many contexts are presented as a device to effectively account for "probabilities" in the decision-making process under conditions of uncertainty.

3.1 On Mind Time: The Sixth Sense of Time

The sense of time is there as soon as any of our other senses are at play. Time is the sixth sense... Eagleman explains this notion of time so well:

> It overlaps the other senses and it is there in the length of a song, the persistence of a scent, the flash of a light bulb (Eagleman 2011).

3.1.1 Stress Triggers and Intuition

In everyday life, you do not remember mundane details such as, what your friend was wearing or what you had for lunch yesterday, unless your friend's dress was

Y. Hayek Kobeissi, *Multifractal Financial Markets*, SpringerBriefs in Finance, DOI: 10.1007/978-1-4614-4490-9_3, © The Author(s) 2013

amazing or the food you had was succulent. What happens is that memory capability improves when emotions are involved. Stress hormones activate the amygdalareceptors in the brain, which modulates the effect of these hormones on the hippocampal consolidation. When the amygdala function[1] is triggered, it influences the encoding and storage of hippocampal-dependent memories, during which more details than usual are assimilated down. The more detailed the memory, the longer the moment seems to have lasted. According to Eagleman, the sense of time is very much related to novelty. Our memory becomes distorted because our brains react more strongly to novelty than to repetition. For instance, when you jump over a river cliff for the first time, the event may seem longer than it actually is, after which you may remember the scenery that you saw in slow motion. When you jump for the second time, your fall will feel shorter and you will not remember the scenery as vividly as the first time. Musser (2011) explains this phenomenon really well:

> When we are sitting through a boring event, it seems to take forever, although when we look back at it, the event is actually not that long. Conversely, when you are doing something exciting, time seems to flash by, even though it may seem longer when you look back and recall the details. In the first scenario, there was little to remember, so your brain collapsed the feeling of duration. In the second scenario, there was so much to remember and as such, the event seemed to expand. In short, time flies when you are having fun, but slows down when you are in quieter, less exciting situations. It is possible that this inverse relation in our perception of time also explains how our experiences shift as we age. Time is different for each observer, where each observer is creating his own universe, his own centre.

We need novelty to stimulate emotions. Whether it is a competition, a threat, falling in love, or any challenge, situations that involve uncertainty serves to trigger our emotions. In the case of the financial markets, there are dull trading times where nothing happens and the day is no different than another. On the other hand, there are highly volatile periods that investors and especially traders never forget, they seem to have lasted longer.

Mind time is how we perceive the passage of time. In *Critique of Pure Reason*, the philosopher, Immanuel Kant (Kant 1787), states:

> Time is nothing but the form of our internal intuition. If we take away from it the special condition of our sensibility the conception of time also vanishes and it inheres not in the objects themselves but solely in the subject or mind which intuits them (p. 33).

In finance, talented professional know how to kick their amygdala into high gear when they have to, while others need an emotional event for this to happen.

---

[1] …your amygdala acts as an emergency control center that gets all the other parts of the brain to quit mucking around with their daily tasks and concentrate all the resources on the one, main thing that is happening. It is like being an athlete or performer "in the zone". When something threatens your life, this area seems to kick into overdrive, recording every last detail of the experience Eagleman (p. 31).

Through experience, skilled professionals have learned how to react to events and figure out precisely what to do next. The process is unconscious; what happens is that through mental effort they acquire enough knowledge (research, analysis…) to attain intuition, an unconscious thinking skill. They can focus and concentrate on the task at hand until they resolve it and they can do that quickly, while others are overcome with panic. Emanuel Derman (Derman 2011) defines intuition[2] very well:

> …In both physics and finance the first major struggle is to gain some intuition about how to proceed; the second struggle is to transform that intuition into something more formulaic, a set of rules anyone can follow, rules that no longer require the original insight itself… Before you can move one level higher in the pyramid of understanding, before you can attain intuition in some domain, you have to struggle with the particulars of that domain until knowledge of its details is second nature to you… Intuition is a merging of the understander with the understood.

In fact skilled traders do not have to mull over a difficulty to resolve it, they do not hesitate. Their hippocampal area triggers the amygdala without the need for too much emotion.[3] This can be explained by the hypothesis that the hippocampal-dependent episodic representation of emotional significance influences the amygdala as well. These types of emotions only need to be imagined and anticipated to create awareness.[4] Experienced traders, through intuition activate their amygdala and concentrate on the situation at hand and its repercussions. For others, when too much emotion is involved; they freeze and panic. Diagram 3.1 highlights how awareness is triggered.

3.1.2 Relative Sense of Time

The notion of time is different for every investor. Different individuals have different priorities and horizons; as such, long-term investors and short-term traders do not have similar time scales. Mandelbrot (2008) concluded that:

---

[2] It is not the intuition as defined by Behaviorist like Kahneman interested in the biases of quick guesses.

[3] Researchers are only beginning to investigate these topics and it is unclear which, if any, of these factors will prove to be important. The relationship between the amygdala and the hippocampus might be bi-directional during the encoding of emotional events. Researchers are just starting to explore more complex interactions between emotion and memory that could be unique to human function (Phelps 2004, pp. 198–202).

[4] The amygdala plays a crucial part in facilitating the link between attention and emotion thus enhancing our memory. For instance, marketing professionals bring into use strategies to trigger attention and curiosity about the uncertainty of what will be said. As a result, the audience's senses will focus on the broadcast commercial/or billboard, and what will be said will not be forgotten easily.

Diagram 3.1   The emotion
cycle

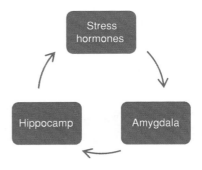

In fractal analysis, time is flexible (p. 238).

In times of speculative excitement, short-term investors perceive the stock prices evolving in a different time scale than calendar time; rather, they perceive the risk and return of a stock in "intrinsic time", a dimensionless time scale that counts the number of trading opportunities that occur, regardless of the calendar time that passes between them. Stocks do not trade continuously; each stock has its own trading patterns. Trading opportunity can be quantified as the rapidity, the liquidity, or the chance to trade a specific underlying (Derman 2002). Volatility clusters and episodes of fast actions intersperse with slow, dull actions.

3.1.3 Mind Gap

The sense of time coordinates all the senses, allowing us to be on top of everything that is happening while our brain recalibrates its expectations. Eagleman's experiments of visual illusions demonstrated that we are living in the past[5]; our consciousness, in fact, lags 80 ms[6] behind actual events. Our brain tries to find/or create an adjusted picture of the world. It gathers up all the evidence from our senses and then reveals it to us. If all our senses are slightly delayed, we have no framework by which to measure a given lag. Reality is carefully censored before it reaches us. We see more quickly than we can hear, but our bodies' process sounds more quickly than light. However, this is not a case for concern as these delays are all synchronized. According to Bilger (2011)

> It's trying to put together the best possible story about what's going on in the world and that takes time.

For instance, senses get processed at different speeds and any differences in the process are erased to achieve an acceptable result. Right now, while Iam typing on

---

[5] According to Libet (2004), "We are not conscious of the actual moment of the present. We are always a little late".

[6] According to Damasio (2002), this lag is actually 120 ms.

Fig. 3.1 Mind Time

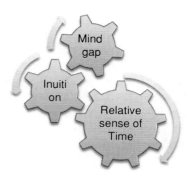

the keyboard, I see what I am typing on the computer screen; I feel the keyboard and hear its clicks all in the same time. We all live in a delayed synchronization. Some facts in nature and events in life are not and cannot be synchronized with our brain (we see the lightning and then hear the thunder). The same mechanism applies to HFT, where the gap for some trades is so wide that other investors get confused by the volume signal. By the time the confused investors/traders get into the market, HFT have already closed their trades (Fig. 3.1).

Figure 3.1 illustrates the concepts of "Intuition", "Relative sense of time", and "mind gap". Being aware of behavior biases and causes, avoiding them and or benefiting from them, is difficult but not impossible if intuition exist. In fact, intuition is the driver of the sense of time; it facilitates the quick interconnection of ideas in the mind. The relative sense of time refers to the way we analyze information based on our time horizon preference. Meanwhile, the mind gap represents one's limit for excelling. In order to develop our analytical capabilities to their highest level, we need to enhance our amygdala awareness and thus our sixth sense. The speed of analysis combined with the ability to interconnect events is vital to successful trading.

3.2 Heuristics

Behavioral psychology studies how people approach problems and makes decisions when faced with complex situations. Unlike the neoclassical school which claims that an individual acts rationally by applying the laws of probability, behavioral psychology stresses the importance of heuristics in decision-making (Kahneman and Tversky 1972). Heuristics refer to the "rules of thumb" we use to make easy and quick decisions. Behavioral finance is the psychology underlying and driving financial decision-making behavior. According to neo-classicists, individuals differ in their actions but have the same risk-averse attitude. Behaviorists, meanwhile, believe that individuals are either risk averse or risk seeking, that although we have the same central tendencies, we differ in our decision-making.

In times of uncertainty, people are overwhelmed with information and tend to use mental shortcuts or heuristics to find quick and easy solutions. This strategy works well most of the time; a professional tennis player, for instance, is likely to beat a novice. However, such simplified strategies can also lead to biases, errors, and confusion in perception, and in investment decision-making. Three major heuristics have been identified by Kahneman and Tversky (Kahneman and Trevsky 1974a, b, pp. 1124–1131) representativeness (84), availability (465), anchoring and adjustment (333). In the following, we explain these heuristics as well as the framing effect.

3.2.1 Representativeness

The representativeness or similarity heuristic is a type of analogy-based reasoning which assumes that similarity in one aspect leads to similarity in other aspects. Tversky and Kahneman (1974) state that in uncertainty, heuristics govern judgment:

> What is the probability that person A (Steve, a very shy and withdrawn man) belongs to group B (librarians) or C (exotic dancers)? In answering such questions, people typically evaluate the probabilities by the degree to which A is representative of B or C (Steve's shyness seems to be more representative for librarians than for exotic dancers) and sometimes neglect base rates (there are far more exotic dancers than librarians in a certain sample).

Indeed, we tend to ignore base rates or the relative frequency with which an event occurs, as well as regression toward the mean (where an extreme value is likely to be followed by one which is much closer to the mean based on underlying factor valuations). This describes the hot hand and gambler fallacy, which state that people make their investing decisions based on a wrongly-assumed:

- Probability of a trend ending—the gambler's fallacy. Where investors fail to understand statistical independence,[7]
- Probability of a trend continuing—the hot hand belief. Where investors pick stocks that were successful in the past, a strategy of short-term momentum continuation. Individual investors who give too much weight to recent performance often succumb to this fallacy.

Investors treat singular data as suggestive and weight it more than the base rate information. The "representativeness effect" makes individuals forecast in a

---

[7] James Bernoulli (1713) developed the law of large numbers: the difference between the probability of an event and the frequency of an event become arbitrarily close to zero as the number of attempts approaches infinity. This means that the ratio of heads to tails will become closer to one after a vast number of flips, not that tails will become more likely after a series of heads. A tail is not more likely on the next flip just because you have just thrown 15 straight heads; the probability of getting another tail is still 50 %.

direction opposite to the base rate trend or the average performance. They often fail to take into account the regression to the mean and the further we divert from the mean, the larger, the error. Singular data usually refers to the most recent information, such as "the market crash of the last six weeks". Shefrin (2011) states:

> We will always rely on heuristics. We will be better but not perfect decision makers. To make behavioral changes you have to put your ego aside.

In reality, market prices can either deviate from or continue their short-term momentum. As such, after a long series of losses, the investor betting against the trend can go bankrupt.

3.2.1.1 Overconfidence Heuristic

Most people who are overconfident about complex tasks have very narrow confidence intervals. They often end up surprised when they do not achieve the expected results and make adjustments very slowly. Washington Post writer, Phillips (1991) cites pilots as an example of overconfidence. After 300 h of flying time, they become so overconfident that their accident rate increases exponentially. A prevalent problem with overconfident people is that they get anchored with their knowledge and experience and do not adjust sufficiently to changing circumstances. According to Russ Fuller (2000), analysts get anchored to their prior descriptive assessments, regardless of whether they are right or wrong. They are overconfident and do not adjust quickly enough when new information comes along:

> People are grossly overconfident regarding their ability and their knowledge. For example, when people say that they are 90 percent sure that an event will happen or that a statement is true, they typically are correct less than 70 per cent of time. Overconfidence can cause investors to under react to new information.

It is very important to be more flexible and not be anchored by our previous practices, especially when they no longer apply to current circumstances. This means that although it is difficult to estimate the proper confidence intervals we need to widen them. In January 2008, an experienced manager about to launch long diversified funds was asked whether it was the right time to invest in the markets. He answered very confidently: "Where do you want the liquidity to go?" Indeed, this manager's belief was confirmed by the market data at that time. The capital of the companies listed in the New York Stock exchange represented more than 100 % of the U.S. GDP, a phenomenon that had not been seen since 1929. In addition, the price earnings ratio surpassed forecasts for the year 2001 by 150 times. Later on, the same year the biggest financial market crash happened and economies tumbled into recession.

3.2.1.2 Valuation Heuristic

Identifying good stocks characterized by low risk and high return, is an example of representativeness. Representativeness involves overreliance on stereotypes. Book-to-market equity and other ratios, such as size, past returns, and past sale growth can be negatively related to expected and realized returns when investors make biased judgments about risk and return in the face of highly-volatile, time-varying, equity premium. The financial soundness of a company is an intrinsic factor. The market price is a result of a number of factors and is estimated as follows:

Price = company quality + stock value

   = intrinsic assessment + external factors differentiating a good company

    from a good stock

Another example of overconfidence is the case of Long-Term Capital Management (LTCM) crisis. The fund managers were so overconfident in their formulas that they underestimated potential investment risks. They convinced investors that markets' inefficiencies are small and temporary in order to be able to use huge leverage. In the beginning of 1998, the portfolio under LTCM's control amounted to well over $100 billion, while net asset value was at around $4 billion. LTCM had become a major supplier of index volatility to investment banks; it was active in mortgage-backed securities and was dabbling in emerging markets such as Russia. When Russia devalued the ruble in 17 August, 1998 and declared a moratorium on 281 billion rubles-worth ($13.5 billion) of its treasury debt, LTCM's equity dropped to $600 million. Banks began to doubt the fund's ability to meet its margin calls but could not move to liquidate for fear that it would precipitate a crisis that would cause huge losses among the fund's counterparties and potentially lead to a systemic crisis.

3.2.2 Availability Heuristic

Some events are easier to remember because they took place recently or were highly emotional. This "availability" heuristic is used to evaluate the frequency or likelihood of an event on the basis of how quickly instances or associations come to mind. Kahneman and Tversky (1974) state that people have a tendency to overestimate the divorce rates if they can quickly find examples of divorced friends or acquaintances. People tend to be biased owing to information that is easier to recall; they are swayed by information that is vivid, well-publicized, recent, or easily retrieved. People estimate the probability or frequency of events based on what they can remember, rather than objective data. The media, meanwhile, uses the availability heuristic to induce people to remember things or events they want them to remember, by making these events more vivid and easier to picture. They steer people away from things they do not want them to remember by making these events unclear and uncomfortable, so that people deny them. Barber and Odean (2006) found that three informational events are especially salient for the average investor: large price movements, quarterly earnings announcements, and news coverage. Unable to evaluate each security based on a deeper analysis, the average investor is likely to purchase securities to which their awareness has been drawn.

3.2.3 Anchoring and Adjustment

The anchoring and adjustment consists of two steps:

- Making an initial judgment or "anchor". Kahneman and Tversky found that when people make a decision, they start from a reference point or familiar position known as the "anchor", regardless of whether the anchor is relevant to the decision.
- Adjusting relative to the starting point. The problem arises when we do not adjust sufficiently from the anchor according to the circumstance.

3.2.4 Framing Effect

The framing effect happens when, in addition to objective considerations, investors' perceptions of risk and return are highly influenced by how they frame decision problems. Variations in the framing of options yield systematically different preferences. For instance, in the context of the housing market, a seller is not concerned by the absolute outcome of his decision (such as selling his apartment for $100,000), but rather by the gains and losses relative to his reference point (selling his apartment for $100,000 when he expected to sell it for $120,000 creates a perceived loss of $20,000) (Paraschiv and Chenavaz 2011). In this example, the seller thinks in relative terms, not absolute terms, and exhibits a decreasing marginal sensitivity. According to the Prospect theory (PT), individuals are risk averse when it comes to gains and risk seeking when it comes to losses, that is:

- people prefer receiving a sure gain; when a decision is framed as a "likely gain", risk averse choices predominate;
- people prefer to avoid a certain loss; where a shift toward risk-seeking behavior occurs when loss aversion manifests itself. An investor's is reluctant to close his trades if doing so would result in a nominal loss. Shefrin and Statman (1985) coined the term disposition effect, as shorthand for the predisposition toward "get-evenitis" (Shefrin 2000). The disposition effect happens when investors

sell winners too early and ride losers too long to avoid regret.[8] The pain decreases after the first time you lose.[9] For investors, the reference point is the purchase price. People are willing to gamble when they are losing, holding on to stocks that have lost value and buying even more to lower their reference price. They realize their gains by selling stocks that have risen in value (Kahneman and Tversky 2000).

- Thaler (Thaler 1999) developed the concept of mental accounting in relation to the framing effect, the decreasing marginal sensitivity and the risk-seeking or loss-aversion biases. Gross (1982) explains how investors create mental accounts with an opening transaction amount and a net current balance:

...the words that I consider to have magical power in the sense that they make for an easier acceptance of a loss are these: Transfer your assets (p. 150).

Investors realize their smaller losses while they continue to hold onto their larger losses. According to Shefrin (2000) these are magical words as they induce the client to use a frame in which he or she reallocates assets from one mental account to another, rather than closing a mental account at a loss. The hedonic editing hypothesis, as introduced by Thaler (1985, 1999) is based on the hypothesis that people have a tendency to frame multiple outcomes in ways that yield the highest perceived value. He states.

One possible place to start in building a model of how people code combinations of events is to assume they do so to make themselves as happy as possible (Thaler 1999, p. 187).

That said, some strategies can be derived to profit from market biases when we are aware of it, e.g.:

- Value and size effect strategy, where we focus on high volume winners and sell-high volume losers. It is important to note that momentum works well among stocks with low analyst coverage[10]; and
- Selecting hot-hand and momentum stocks at the same time while paying attention to market timing and reversals.

While it is important to learn from experience, we should be mindful of factors that seemingly do not influence our decision-making but in fact do. One way to achieve this is to supplement heuristics with objective information such as reliable information. People will always be affected by heuristics and consequently arrive at irrational decisions. The trick is to play "offence", that is, if something seems right, look for contrary evidence that will validate your decision before putting that decision to work.

As new information is constantly entering the market, financial participants revise their expectations according to their own utility perception. As such the study

---

[8] except during the month of December in the U.S., for tax reasons.

[9] For the decision maker, each additional monetary unit gained or lost is worth less than the previously gained or lost monetary unit (Kahneman and Tversky 1992).

[10] High volume stocks are generally *glamour* stocks and low volume stocks are generally *value* or *neglected* stocks, (Lee and Swaminathan 2000).

of utility is important to understand the financial marketplace. The key element in the information content is the surprise element. Surprise is experienced only if the outcome occurs and so is derived its utility per individual. Bearing in mind that information[11] is a decreasing function of probability, below we introduce an innovative subjective utility theory based on Viole and Nawrocki utility functions.

3.3 Multiple Heterogeneous Benchmark Utility Functions

As partial moments allow for different targets to be calculated with variations in degrees; they are highly configurable to multiple constraints and do not require any distributional assumptions. The separate partial moments also allows for considerable asymmetry of the curve from the delineation of negative to positive utility. Viole and Nawrocki extend the upper partial moment/lower partial moment (UPM/LPM) model of Holthausen D (Holthausen 1981) to include a multiple UPM/LPM model that contains both positive and negative utility considerations with single and multiple targets. The formulas representing the n-degree LPM and q-degree UPM are:

$$\text{LPM}(n, h, i) = \frac{1}{T}\left[\sum_{t=1}^{T} \max(0, h - R_{i,t})^n\right] \tag{3.1}$$

$$\text{LPM}(q, 1, i) = \frac{1}{T}\left[\sum_{t=1}^{T} \max(0, R_{i,t} - 1)^q\right] \tag{3.2}$$

where $R_{i,t}$ represents the return of the investment i at time t,
h is the target for computing below target returns, and l is the target for computing above target returns.
n is the degree of the LPM, q is the degree of the UPM (Table 3.1).

---

[11] In 1952 Shackle (GLS 1952) advanced that an occurrence with low probability contains more surprise than an occurrence with a high probability with the amount of surprise measuring the risk of the investment. As such information is a decreasing function of the probability. So, if there are different weights of outcomes (potential surprise), then there will be a *potential surprise function*. The *potential surprise function* of Shackle is analogous to the weighted entropy measure derived by Guiasu (Guiasu 1977). Bill Harding and N proposed a state-value weighted entropy value 25 years ago because entropy as a statistical measure does not take into account the values of the microstates (Nawrocki and Harding 1986). Weighted entropy in comparison is more adequate the economic and financial context as a measure of portfolio risk, because the structure of the dispersion contained in the frequency classes is not ignored.

$Hw = -\sum_{i=1}^{n} Xi\, pi\, \log_e pi$
Where **Hw** is the weighted entropy,
Xi is the monetary payoff or return,
n is the number of outcomes, and
pi is the a priori probability of the outcome **I**.

Exponents' n and q will represent the investor's sensitivity to losses and gains respectively as identified in Eqs. 3.1 and 3.2 above. These exponents will serve to augment any curve characteristics.

By recreating the Allais paradox (Allais 1953) and the (PT) question set (Kahneman and Tversky 1992), Viole, F, and Nawrocki, D (Viole and Nawrocki 2011) found subjective wealth to be the major determinant of risk aversion and were able to reconstruct a fourfold utility function demonstrating both PT and expected utility theory (EUT) are quiet relevant.[12] The major difference in functions rests in the points that:

1. Individual make their decision based on change in wealth rather than total wealth. However, if identical amounts were offered to a nominally wealthy and nominally less wealthy individual, the change would be a reflection of the nominal wealth (Viole and Nawrocki 2012).
2. There are two reference points, not one per EUT and PT, in the value function, an upside target for the gain function (e.g. limit order), and an acceptable level of loss (e.g. stop loss) for the loss function.
3. For aggregate wealth the benchmarks are the subsistence level (S) for losses and the personal consumption point (PCS) for gains.
4. There is a difference between tangible and intangible utility.
5. Individual make their decision based on subjective wealth qualification. Subjective wealth creates the reference for the decision which is based on both absolute amounts and relative amounts (PT and EUT respectively).
6. Individuals are loss averse when it comes to gains approaching toward their PCS levels.
7. But Gains are not ultimately concave. Above the PCS level: the intangible quality of money encourage a gain seeking attitude.
8. Individuals are loss seeking when it comes to losses.
9. But losses are not ultimately convex. Below S: loss aversion behavior is pronounced.

To find out the final decision we compare focus gain and focus loss to S and PCS:

- Focus loss and focus gain are expectations of an external event (market returns).
- S and PCS are internal individual limits of aggregate wealth to have some effect on the decision maker for local investment decisions.

---

[12] According to the Expected Utility Theory (EUT) the decision maker chooses between risky or uncertain prospects by comparing their expected utility values, i.e., the weighted sums obtained by adding the utility values of outcomes multiplied by their respective probabilities. The cumulative prospect theory of Kahneman and Tversky's (Kahneman and Tversky 1992) is a descriptive theory of decision behavior where weights are applied to the cumulative probability distribution function and relative value is assigned to each outcome. In this theory, the concept of "utility" is replaced with the concept of "value". The reference points become the net gains and losses instead of net wealth. The prospect theory is characterized by a value function that is concave for gains, convex for losses, and steeper for losses than gains.

Table 3.1 UPM/LPM utility estimation

$U(x)$	LPM Conditional Values
$f(-\text{LPM}(n,h,x) - \text{LPM}(n,y,x) + U(0))$	If $\text{LPM}(n,h,x) \geq \text{LPM}(n,h,y) > 0$
$f(-\text{LPM}(n,h,y) + \text{UPM}(n,y,x) + U(0))$	If $\text{LPM}(n,h,y) \geq \text{LPM}(n,h,x) > 0$
$f(-\text{LPM}(n,h,y))$	If $\text{LPM}(n,h,x) = 0$
$U(x)=$	UPM Conditional Values
$f(\text{UPM}(q,l,a) - \text{LPM}(q,a,x) + U(0))$	If $\text{UPM}(q,l,a) \geq \text{UPM}(q,l,x) > 0$
$f(\text{UPM}(q,l,x) + \text{UPM}(q,a,x) + U(0))$	If $\text{UPM}(q,l,x) \geq \text{UPM}(q,l,a) > 0$
$f(\text{UPM}(q,l,a))$	If $\text{UPM}(q,l,x) = 0$

UPM () = Upper partial moment (degree, target, benchmark)
LPM () = Lower partial moment (degree, target, benchmark)
n = Investor's loss aversion level
q = Investor's gain seeking appetite
h = Target for computing LPM
l = Target for computing UPM
x = Investment
y = Benchmark Y
a = Benchmark A

- The historical benchmark is the "status quo" or customary wealth noted at the 0 point on the x-axis. The concave utility for gains from the (0, 0) point and convex utility into the S level reflects this notion.
- For aggregate subjective wealth within our internal benchmark (S, 0; PCS) we portray a PT behavior.
- With local investments or aggregate subjective wealth Beyond our internal benchmarks we portray an MSD behavior (Fig. 3.2):

The fourfold utility function explained:

1. At (0, 0): Information surprise effect is too high. The information content is maximal. We are far from our maximum utility satisfaction or satiation level. The probability of reaching it is so low; the surprise content is very high if the expected return is realized. We do not know when and how the expected return will reach our satiation level or max utility. It can be translated into Max entropy at the market level as for sure all participants would not be satisfied with a 0 utility. So, low probability means low utility and max surprise if the event occurs.

2. (0, 0; PCS) Afraid of the potential negative surprise of moving too far away from our internal satiation level, we are loss averse and try to stay as much as possible in the positive side. The aim is to carefully stay far from (0, 0). Risk aversion dominates.[13] As we are in our comfort zone we long to stay there as much as possible and do not react immediately to news. If PCS is the min entropy level, then once we reach it, a bifurcation will happen or will go beyond it for the individual depending on exogenous market returns. When returns

---

[13] …The less wealthy is typically guided by the concavity of the total wealth function for local choices, as they are furthest from their PCS level (Viole and Nawrocki 2012).

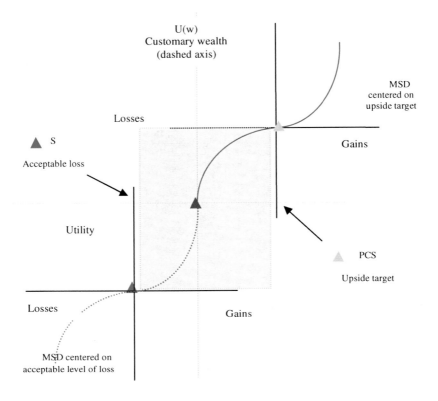

The fourfold utility function

The shaded area replicates the twofold Prospect theory value function (Viole & Nawrocki, An analysis of heterogeneous utility benchmarks in a zero return environment (2011b), 2011). Our curve is simply two Markowitz Stochastic Dominance (MSD) Reverse-S curves that meet at customary wealth; one curve centered on an upside target, and the other curve centered on an acceptable level of loss. In the aggregate wealth function, the upside benchmark is the PCS and the lower benchmark is S, Roy's (1952) safety first level of subsistence. Superimposing 2 MSDs onto larger utility axes reveal final shape of our function: concave - convex – concave – convex.

▲(0, 0) correspond to maximum entropy.
S is the subsistence level for aggregate wealth, acceptable level of loss for local investment decision. Bifurcation point at min entropy.
PCS is the personal consumption satiation point for aggregate wealth. Upside target for local investment decision. Bifurcation point at min. entropy.

Fig. 3.2 Multiple Heterogeneous Benchmark Utility Functions

expectation (focus gain) approaches the satiation level, the information content diminishes as the probability of reaching it becomes so probable (probability is high): What happens to the utility value? It does not change. We still want to get our satisfaction. This is a subjective individual satisfaction and utility is still high. Once we reach it, we might end up in three scenarios:

- (Above the PCS level), Future expectation (focus gain) might change. And go higher than the satiation level. We become gain seekers,[14] because if we lose; we have enough margins to only go back close to the already satisfying PCS level.[15] Money has an intangible utility and "the sky is the limit". The bubble is building up/benchmarks have been exceeded. Any market correction that gets us back to the PCS level will ensure a bifurcation (horizons alignments, maximum consensus, maximum Rho), and a return to (0, 0) or max entropy to adjust markets gaps.

- Future expectation does not change and we enjoy our position. The probability of the focus gain $p(x)$ being reached is now 1, minimum entropy point at the individual level.

- If the sum of individual satisfaction level of all market participant (considering their weight) is reached: Any negative news from the market will affect them as it will have a high content of information or surprise. By trying to realize their gain altogether at the same time (correlated horizons), it triggers the bifurcation phenomenon to assimilate the information. Bifurcation being a market event triggered by the global action of individual forces, this is where the microstate weighting comes into play. If all microstates had equal weighting then yes, it would depend solely on number of individuals. However, as explained by Viole (Viole 2012) if a heavily weighted microstate enters, say the Fed, then minimum entropy can be achieved regardless of the number of individuals. In essence that is the role of the Fed, to alleviate uncertainty—reduce entropy. They just call it price stability and full employment. The fed action is a market factor creating information surprise of different amplitudes. If the fed action was expected then it will reassure the market and reduce entropy or, if the action was unexpected, information surprise creates a shock. Mean reversion process follows.

---

[14] …When wealth increases and the PCS level is eclipsed the Wealthy enjoy a convex utility or economies of scale for additional resources. The migration of positioning explains the non-stationarity of loss-aversion as the marginal dollar gained is larger than the marginal dollar lost for the Wealthy. Again, the total subjective wealth influences the change in wealth of the decision; neither is exclusive per EUT and PT (Viole and Nawrocki 2012).

[15] The house money effect introduced by Thaler and Johnson (1990) predicts investors will be more likely to purchase risky stocks after closing out a profitable trade. It is an example of mental accounting, whereby agents consider large or unexpected wealth gains to be distinct from the rest of their wealth, thus they are more willing to gamble with such gains than they ordinarily would be.

3. (Below our S), we act as if we wanted to go back to it and not lose anymore. First we focus on getting to our S, then we will focus on getting to (0, 0). Fear impedes us from reacting immediately to information, at least not before second thoughts. Risk aversion with losses is the motto.
4. (S; 0,0) Within this bracket we are very sensitive to losses and we focus on avoiding it by doing whatever it takes (overconfidence): risk seeking is worth it. The focus is getting out of this loss zone as quickly as possible which explains why the utility curve becomes convex, trying to reach the breakeven point (0, 0). So bifurcation happens at S (min entropy) to reach optimal entropy. Thus ensuring survival and start all over again!

3.4 Fuzzy Logic and Bayes' Theorem

Fuzzy logic and Bayes theorem are different notions that showcase the weakness of probability calculations when subjective parameters are at play. When you read this section, you will learn how individuals classify situations especially under uncertainty and how to account for prior conditions in the probabilities calculations. It is however, delicate to state that these notions are ultimate.

3.4.1 Fuzzy Logic

According to the fuzzy logic, we often get confused between probabilities and heuristics and end up applying the principles of fuzzy logic. Individuals classify objects in order to identify them. This classification process involves identifying similarities between the objects or the situations observed and the objects or situations retained in our memory. In a classic binary grouping, an element has a value equal to zero or one, that is, an element either belongs or does not belong to a group. The problem arises when classifying an element that belongs partially to a group. The classic concept of grouping does cover sub or partial grouping; as such, Zadeh (1965) introduced the concept of the "fuzzy group," which addresses more complex grouping scenarios such as partial belonging. In fuzzy logic, we can rate an object between zero and one in order to describe the degree to which it is more similar to one group than another. For instance, if X is a reference group with several classes, then each class is a sub-group of X. If A is one such sub-group, it is defined by a characteristic function $x_A = 0$ for the elements of X not belonging to A and $x_A = 1$ for the elements of X belonging to A (Yudknowsky 2003).

That said, not all subgroups are clearly defined. A fuzzy subgroup of X, for instance, is defined by the function of belonging that associates with each element x of X, the degree $0 \leq f_A(x) \leq 1$, where

$$x_A : X \rightarrow \{0; 1\} \quad f_A : X \rightarrow [0; 1]$$

Thus, a classic subgroup is a particular case of the fuzzy subgroups. The two extreme cases of fuzzy X subgroups occur when X itself is associated with a function that is equal to one for each element of X; and $\emptyset$ is associated with $f_x \emptyset$ for each element of X. The general rules of fuzzy logic are summarized as follows:

A = degree of belonging to group A, where $0 \leq A \leq 1$
$1 - A$ = complement of A
Min (A, B) = intersection of groups A and B
Max (A, B) = union of groups A and B

Fuzzy logic allows the concept of grouping to be applied to situations where the concept of subgrouping is very different. It is important to note, however, that an element can belong to a subgroup and its complement at the same time, with the total degrees of belonging equaling to one. The concepts of fuzzy groups and probabilities are different, as each one measures a different aspect of the level of uncertainty in a group. The function of belonging is a function that allows a complementary state to be described. If an element is similar to an ideal element, its degree of belonging is equal to one. As a consequence, the addition of new elements does not change this element's function of belonging. The probability function, on the other hand, changes with the addition of new elements because it depends on the frequencies of the elements that it consists of. In this regard, it is useful to recall Kahneman and Tversky observation that certain questions—such as whether federal authorities will raise interest rates in the next few months, or whether a recession is about to occur—do not have precise responses. A 70 % chance of recession indicates that the degree of belonging of the present financial and economic conditions to "recession" is 0.7. Analysts also base the rationale for their opinions on fuzzy logic. For instance, if an analyst considers a price earnings ratio, PER = 25 as a point of reference for an equity belonging to the group "equities to buy", an equity with PER = 30 will be classified as "equities to buy with moderation" because its degree of belonging to the first group is 60 %. In sum, a fuzzy set is one that requires fewer rules and variables and uses linguistic rather than numerical description.

3.4.2 Bayes' Theorem

Bayes theorem (Bayes 1763), also known as the inverse probability law, describes how to change prior beliefs that we might hold about hypothetical events into new, posterior, beliefs given evidence about how likely the events are. As such it provides a mechanism to update our prior knowledge based on new data or observed evidence that might give greater support to one event over another. It also works in the inverse situation: given some posterior facts we can determine the prior beliefs most consistent with those facts and it is this ability to reason from cause to effect and vice versa that makes Bayesian reasoning powerful. The theorem relates the probability of the occurrence of an event to the occurrence or nonoccurrence of an associated event. Although it can apply to more than two events, we will explain the theorem using only two events for simplicity:

A is the already known (basis hypothesis)	A = Market is up
	(~A) = Market is down
X is the implication	Positive news
Prior probability	p(A) = 50 %

Conditional probability (1): Positive news given market up (X|A) = 70 %

The reading is from right to left i.e. the degree to which the market is up implies the news is positive. The probability for X given A is the degree to which A implies X, while the probability of X given ~A is the degree to which ~A implies X

Conditional probability (2): Positive news given market down p(X|~A) = 20 %

Or the degree to which the market is down given news is positive.

p(X|A) + p(X|~A) does not have to add up to one. We still have other probabilities where A* implies X. For example, the market is given up no specific news where p(X|A*) = 5 %

The posterior probability of these two events is:

$$p(A|X) = \frac{p(X|A) \times p(A)}{p(X|A) \times p(A) + p(X|\sim A) \times p(\sim A)}$$
$$= \frac{0.7 \times 0.5}{(0.7 \times 0.5 + 0.2 \times 0.5)} = 0.8$$

People do not take prior frequencies sufficiently into account, for example, there is a 70 % chance that the market will go up if we receive good news, then this kind of reasoning is insensitive to the prior frequency given in the problem; it does not notice whether 50 % of news or 10 % of news is good.

Appendix A: NASDAQ100 (2000–2001)

In this section, we analyze the behavior of investors from 19 October 1999–12 January 2001, to demonstrate the biases in the markets, using the Nasdaq 100 (NDX100) and the S&P500 volatility index (VIX). Figure 3.3 illustrates the

Fig. 3.3 The NASDAQ 100 Index from 19 October, 1999–12, January, 2001. (*Data Source* Bloomberg)

movement of these two indices during the period of our analysis in three phases: euphoria, transition, and bear market.

Euphoria

The collective myopia observed during this phase was a logical consequence of individual investors' irrational behavior where more importance is given to information that satisfies their desire for maximum profits.

During the accumulation phase (19 October, 1999–3 January, 2000), prices started to increase progressively, influencing the perception of market players, leading to widespread optimism. The NDX100 rocketed from 2,362 points to 3,790 points, an increase of 60.45 % in only 53 working days. Everything in the environment was functioning in favor of sustainable growth as evidenced by high liquidity, favorable forecasts for companies, and prospects for economic growth combined with limited t inflationary pressures. The increase observed during this phase was accentuated by investors ready to repurchase the shares they had sold at lower prices as soon as there was a market correction. No one wanted to or could afford to miss the market rise. Trading in the market became a social trend, a fashion, allowing individuals to show off their wealth, and prove their skills. Thereby they were easily induced into the leverage buildup cycle.

The period between 4 January 2000 and 10 March, 2000, was marked by technical corrections and lots of speculation but overconfident behavior reinforced the bullish trend. Throughout January 2000, the NDX100 oscillated between 3,790 and 3,446 points, a fairly reasonable difference of 344 points or 9 %, owing to investors' profit-taking. Investors created mental accounts, closing profitable positions and leaving open losing ones. Their unrealized losses were only considered an accounting loss or "paper loss" as long as they were not forced to realize them, such as in the case of a margin call.

On February 2 2000, the Federal authorities increased the rates by 25 basis points (bps) instead of the expected 50 bps, thus reinforcing the positive bias. Markets soared and on March 10, the NDX100 reached 4,587 points, an increase of 33 % in only 28 days. Investors became even more overconfident and increased their investments through leveraging. In fact, decisions were biased by the information selected to fit their thoughts and justify their behavior, otherwise known as anchoring behavior. Investors ignored cautionary calls by analysts who foresaw possibilities of market deterioration because of their desire for rapid gains and fear of missing a market upturn. They continued to leverage their positions, consequently amplifying the difference between market prices and fundamental values.

During periods of euphoria, the sensitivity of market players to media is incredibly high; an example of this phenomenon is illustrated by the so-called "Media effect". Trading rooms are often tuned into financial media channels, where traders listen to the opinions of analysts. While TV channels may not be the best source of information, it is the most accessible to the trading public. Price fluctuations accompanied by the recommendations of financial journalists, increase (or decrease) the number of stock market transactions exponentially. It is thus useful for professionals to study how media influences the general public in their decision making.[16] The Media effect takes place when investors follow market tidbits or snippets of information. This effect is especially pronounced in periods of market excitement during which the categories of investors are quite diverse, ranging from large fund managers and professional investors, to nonprofessional investors and retired individuals. The following is a non-exhaustive list of psychological mechanisms by which journalists, intentionally or unintentionally influence their audience:

---

[16] Keynes described the positive feedback loop between professionals and the general public in his well-known metaphor about a beauty contest in which he compared the stock market with the competition between American newspapers where each competitor tries to select the photo they think would appeal to the average America: "The question is not to choose, according to your own opinion, the photos that are actually the prettiest, nor to select the one that the average American would consider the prettiest. We have reached here the third degree where we consecrate our intelligence to anticipating what the general public thinks to be the opinion of the general public" (Keynes 1936, p. 156).

- Financial journalists identify with certain social groups and as such, they are more likely to influence those groups. For instance, by writing an optimistic article during a bullish phase, journalists influence their readers to increase their investments. Tvede (2002, pp. 191–192) explains this effect using a medical example test.

  Imagine you are presented with the following medical outcomes:

  – Outcome One: 200 people out of 600 will be saved whatever option is selected.
  – Outcome Two: 400 people out of 600 will die whatever option is selected. The result of the test is that when people are presented with the first outcome, they choose an option; but, when presented with the second outcome, they fail to make a choice.

- Journalists often ignore criticism that calls into question their previous analyses. They tend to wrongly interpret new information in order to confirm their opinions.

Between 13 March, 2000 and 7 April, 2000, economic indicators confirmed the build-up of an economic heating and the future deterioration of the financial environment. Investors began to fear the possibility of a 50 bps increase in the interest rates during an upcoming federal meeting on 21 March. As a wave of anxiety started to sweep investors, the NDX100 fell by an average of 3 % per day between 13 and 15 March. On 21 March, the federal authorities increased the interest rates by only 0.25 point, which somewhat calmed the market. As the outlook turned positive, the market recovered. Here, we can discern the hesitation of the investors prior to the market reversal. They have a selective perception, unconsciously interpreting information incorrectly to rationalize their strategies and a selective exposure, where they are open only to information which validates their outlook. Unconsciously, investors adopt the same attitudes of others they identify with. However, they overestimate the number of people sharing their opinion. Meanwhile, large fund managers began to pull out of the market.[17] The deterioration of the financial market originally forecast becomes a reality. Between 28 and 30 March, the NDX100 fell by 10 %. Both investors and the media attributed this decrease to a simple market correction owing to profit-taking. On 31 March, the market rocketed to 4,397 points.

On 3 April, the market tumbled again, this time by 7.6 %. On 4 April, it fell by 13 % and closed at 4,034 points. The market bounced back on 7 April, closing at 4,291, an increase of 4.9 %; investors breathed a sigh of relief as the volatile week ended positively. It was still unclear, however, whether the market correction was simply a technical correction. Nonetheless, large investors had already started liquidating their positions.

---

[17] An example is Goldman Sachs chief market strategist Abby Cohen's cautionary warnings on 28 March 2000.

Transition phase—the crash

The process of financial disengagement started as signs of a reversal grew stronger, monetary conditions tightened and anxiety took over. This period can be compared to trying to listening to an opera. We are so carried away by the music that we are able to block out any disturbing noise. As time passes, however, when the ambient noise intensifies, we can no longer ignore it and continue to enjoy the opera. This increasing noise represents the financial disengagement of market professional. Investors realized that the correction observed in the stock market prices was not the same as the previous one. This correction was more serious and did not seem selective and brief as it affected all market sectors. On 10 April, the NDX100 fell 300 points and prices continued to decline drastically in the following 4 days prompting authorities to suspend quotes on several shares. The successive suspension of quotes, however, aggravated the situation even more, owing to the temporary illiquidity it created and the following behaviors were observed:

Crowd behavior: the panic sent a clear signal of the critical situation to other market players, influencing the rest of the investors who ended up changing their attitudes. In short, the panic was generalized with optimal dissemination of information and investor's horizons align. Many investors habitually adopt strategies that take into account long-term equilibrium. From time to time, however, investors forgo these strategies when they lose confidence in the market and in its future. The lack of comprehension of external events provokes a panic more accurately described as an "avalanche effect".

Sensory-tonic theory: The pressure created metabolic reactions which reinforced the escalation of panic. Subjected to stress, investors started to liquidate their positions. Liquidation made market prices fall even more and increased the margins call even more.

The market continued to drop every minute without showing any signs of stopping; on 14 April, the NDX100 closed at 3,205 points, a decline of 1,086 points (23 %) in 5 days. The Bearish trend (17 April–31 May, 2000) that followed the crash was marked by erratic fluctuations. No consensus could be reached and market players were lost. The mood fluctuated between positive and negative depending on market news. As such, investors were unable to think clearly and stick to their decisions without first having to review and analyze their decisions several times, ruminating over and over again. Now, the only pertinent information was the negative yields of the prices. In addition, margin calls amplified daily volumes on the sell side and aggravated the bearish trend.

On 16 May, indicators continued to show signs of economic heating, prompting federal authorities to increase interest rates by 50 bps. As a result, the market tumbled by 600 points. Owing to this substantial fall, investors hesitated for several days before buying in the market again. On 30 May, the NDX100 climbed by 9.6 % to 3,414 points. During this period, market fear was reflected in the high implied volatility where the VIX Index fluctuated between 27 and 35 %. Investors questioned whether this volatility spike indicated a short-term correction or if it was the beginning of a more severe correction. In reality, investors were hoping for a technical rebound following news that inflation has not reached an alarming level and that the economy was going to witness a soft landing. That was followed by a consolidation phase between June and September 2000 where the NDX100 fluctuated between 3,477 and 4,099 points.

Uncertain investors remained on their guard and the market remained nervous until the process of elimination started at the end of September 2000. Any negative information, regardless of its importance and whether it concerned only one company, affected the whole sector. The progressive elimination of categories of investors continued with the margin calls. During this period, it was crucial to ensure that the decrease in the price per barrel of oil was definitive and to evaluate the consequences of the fall of the Euro, which created an uncertainty relative to the period of profit reporting. The period of pre-announcement of earnings or warnings was an excuse for everyone to liquidate their positions. The NDX100 lost 13 % in 20 days. Investors were obsessed with the slightest details and were questioning themselves whether they had run proper analyses and whether they had overlooked any important information. They ended up by developing symptoms of depression in the sense that they were often unsatisfied and preoccupied. On October 2, 2000, during the Federal Reserve board meeting, Chairman Alan Greenspan's commentary was pessimistic, owing to the resurgence of inflation fed by the increase in the oil price. Simultaneously, the weakness of the Euro against the U.S. dollar affected multinational's margins which led to a series of "profit warnings" on behalf of these multinationals.

Following these events, an inverse schema was rapidly established wherein the decrease in share prices, deterioration of the underlying mechanism and bias reinforced one another. The NDX100 was in a bearish pattern (2 October, 2000–12 January, 2001) and the consensus was mostly negative. Market participants waited for a republican candidate to be elected, hoping for a new flow of liquidity owing to substantial tax reduction. By November 8, the elections were still in progress and the market was overwhelmed by enormous uncertainty pushing the NDX100 down by 7 %. At the end, G.W. Bush was elected as U.S. President on December 13. By then, however, pessimism had already overtaken the market. Market players felt that only a decrease in interest rates by the Federal Reserve could save these companies and uplift market, but nothing happened. Therefore, the NDX100 declined by 5.8 % the next day, to 2,340, a decrease of 34 % in 63 days.

On 3 January, following another 50 bps drop in interest rates, investors were very pessimistic and overestimated the deterioration of the economy and the markets. They were influenced by information, including that which was irrelevant, a phenomenon also known as the "touchy-feely syndrome." The NDX100 declined by 48 % and each market rally became a "bear trap". Throughout the bear market, an increase was deemed convincing only if the advance was based on economic and financial fundamentals news which can only modify the underlying mechanism.

Chapter 4
Macrostates Indicators

Abstract Economic cycles are the strange attractor of the financial process. Economic indicators and their inter-relationships with each other are reviewed in this chapter. The specifics of each cycle make it unique; economic indicators by themselves are not as valuable, it is their causes and the magnitude of their effects that provide value and insight. With this interconnectedness in mind, we are enabled to explore ways to perceive the cycle ahead. Since financial markets strive to foresee economic fluctuations, it is crucial to be able to identify the characteristics of the economic cycles and to analyze those specific characteristics. The identification of macrostate factors (economic cycles, disequilibrium, and changes) of market's characteristics and the accurate evaluation of the sensitivities of a portfolio to those same risk factors are fundamental tools for all investors wanting to guard against the sudden reversals of trends.

4.1 Delayed Response Mechanism

Nonlinearity of cycles explains why it is so tricky to predict financial markets. Cycles, generated by nonlinear, delayed effects, do not recur in exact fashion due to changes in initial conditions. In order to determine the next cyclical behavior, we need to identify the delayed response mechanisms involved (a short list is presented in Table 4.1) and use them as inputs in our forecasting model.

Schumpeter (1939) developed a global theory on the cyclical movements of the economy and the financial markets. In his model, three main events played a significant role in the markets: investments in the production of goods; company shares; and the growth of credit. The real world, however, is more complex than Schumpeter's model—economic scenarios can change radically especially with the occurrence of unpredictable events. Butterfly effects for instance, can occur in the economy. In addition, economies are regulated by underlying cyclical movements which mutually superimpose themselves on the markets and whose frequency cannot be predicted. Because of the butterfly effects and the bifurcations

Y. Hayek Kobeissi, *Multifractal Financial Markets*, SpringerBriefs in Finance, DOI: 10.1007/978-1-4614-4490-9_4, © The Author(s) 2013

Table 4.1 Short list of delayed response mechanisms

Cause, change in macrostates	Effect, change in	Delay
Interest rate changes	Capital spending	6–18 months
Birth rate	Birth rate	20–30 years
Business profitability	Production capacity	1–2 years
Natural resources prospecting	Production	3–10 years
Stock exchange prices	Real estate prices	6–12 months
Employment	Consumer spending	3–6 months
Price	Demand	0–24 months
Salaries of a category of workers	Availability of such workers	3–8 years
Credit availability	Number of bad loans	1–5 years
Money stock	Inflation	6–18 months

Source Triple.net

caused by the feedback loops, these cycles become limited but non-periodic and they do not resemble each other. Analyzing economic cycles is important as it helps identify the general characteristics of the market under study. Economic indicators and their interrelationships with each other shed light on a market and its circumstances. The specifics of each cycle make it unique; economic indicators by themselves do not make them valuable, but rather their trends and repercussions do. However, every indicator serves as an important signal which can help us identify trend reversals and the specifics of a cycle.

Famous economist Fisher I. (Fisher 1933), developed the debt deflation theory of great depressions where in booms and depressions, at some point in time, a state of over-indebtedness exists (if no external action is taken) leading to liquidation through the alarm either of debtors or creditors or both. The following table as explained by Fisher gives a fairly typical, picture of the cross-currents of a depression in the approximate order in which it is believed they occur.

Leading indicators specify in advance the economic activities and their outcomes (Diagram 4.1). Economic activities influence market movements and the behavior of each sector in relation to one another. Figure 4.1[1] shows the composite index of the 10 leading indicators which include: the average work week for manufacturing industries; number of people declaring unemployment and claiming employment insurance; index of vendor performance[2]; S&P 500 index; monthly

---

[1] The Economic Research Institute Weekly Leading Index (ECRIWLI) is promptly made available to the public and is updated on Fridays at 10.30 a.m., EST. This index addresses a number of limitations in the widely reported leading economic index (LEI) originally developed by ECRI's founder, Geoffrey H. Moore, for the U.S. Department of Commerce. Every Friday, the index is updated with data from the previous week. This allows a significant and timely monitoring of economic conditions.

[2] This index calculates the percentage of companies indicating a slowdown in deliveries. Generally, an increase in the index translates into an economy at full growth, while a decrease in the index translates into an economy that has slowed down where producers can rapidly carry out deliveries, as they are not overloaded.

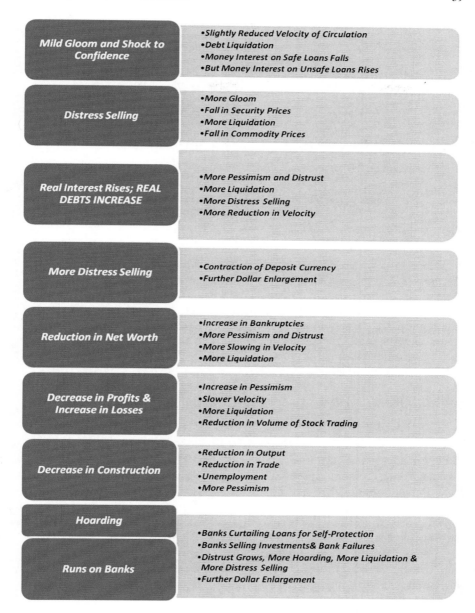

Diagram 4.1 Fisher's nine factors, recurring, and occurring (together with distressed selling)

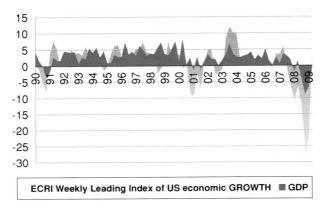

Fig. 4.1 Composite index of leading indicators. *Source* Bloomberg

adjusted liquidity (M2 adjusted)[3]; index of new construction permits; new orders
for consumer goods and equipment; contracts and factory orders; consumer sentiment[4]; the spread between 10-year treasury bonds; and the rate of federal funds.

Since 1959, the composite index, on average, has predicted a recession 2
months in advance and an economic peak 8 months in advance. The following
figure shows that the composite index reached an all-time low in 2008, just before
the markets collapsed.

4.2 Equity Market Indices, Consumer Confidence, and Unemployment

Equity market indices determine the engagement or disengagement of a category
of financial agents with respect to economic and financial forecasts. Although it is
clear that financial markets act in advance, we cannot predict the exact timing of
actions. Thus, if financial disengagement is caused by expectations of an economic
slowdown, it will only serve to amplify the economic deterioration that follows,
negatively affecting the wealth of the economy. Previous economic crises have all
been preceded by a fall in consumer confidence, followed by deterioration in the
employment market and the advent of widespread company layoffs, as illustrated
in Figs. 4.2 and 4.3, respectively.

---

[3] Observing liquidity mainly consists of (1) following the composite indices of leading
indicators for the absorption of the liquidity for productive means, such as investments, new
industrial orders, and number of construction permits; (2) establishing forecasts on the deposits
and withdrawals made by central governments; and (3) following the markets for interest rates,
the rate curve, and the price of gold in the market.

[4] The consumer confidenceindex indicates how likely consumers are to make favorable decisions
to buy durable and nondurable goods, services and homes.

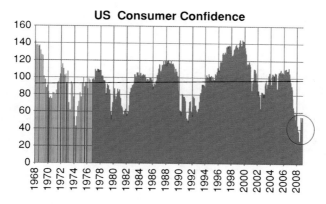

Fig. 4.2 Decline in consumer confidence. *Source* Bloomberg

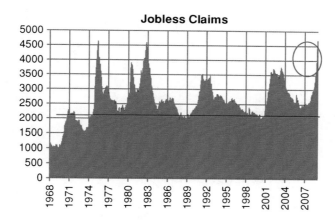

Fig. 4.3 Surge in unemployment. *Source* Bloomberg

These observations can be explained by analyzing the credit conditions during disengagement which reveals a reduction in credit for companies, as illustrated in Fig. 4.4.[5] Companies find themselves facing two constraints: a decrease in

---

[5] The *Federal Senior Loan Officer Opinion Survey on Bank Lending Practices at Selected Large Banks in the United States* is a quarterly survey of approximately 60 large domestic banks and 24 U.S. branches and agencies of foreign banks. The survey questions cover changes in the standards and terms of the banks' lending and the state of business and household demand for loans. The survey often includes questions on one or two other topics of current interest. The Bloomberg Financial Conditions Index combines yield spreads and indices from the money markets, equity markets, and bond markets into a normalized index. The values of this index are z-scores, which represent the number of standard deviations current financial conditions lie above or below the average during the June 1994–2008 period. A full explanation of the methodology can be found in the Financial Conditions Watch help page, available at Bloomberg {NSN KX0MOS3PWT1C <GO>}.

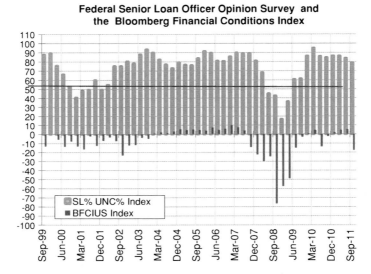

Fig. 4.4 U.S. credit conditions and the financial markets. *Source* Bloomberg

consumption and a simultaneous need for working capital. They end up adopting a cost-reduction strategy starting with layoffs. This obviously, has socio-economic repercussions and amplifies the crisis.

4.3 Leverage and Market Liquidity

High leverage encourages excessive risk taking. And when it is identified it translates the overconfidence and optimism prevailing in the markets leading undoubtedly to systemic risks. The 2008 marked the collapse of the highly leveraged financial entities that pumped money into the housing sector through overleveraged virtually "un-hedgeable" opaque derivative structures. As such, tracking leverage indicators are vital, the issue is that finding ways to estimate and track leverage is not as straightforward as it seems. For instance, finding this information is hard in opaque structures.

Governments hold substantial influence over economies and markets and as such, the prevailing perception is that it is best to monitor fiscal policies. To control economic growth, central banks have at their disposal many instruments: buying or selling national debt; changing credit restrictions; and changing the interest rates by changing the reserve requirements to name a few. Since 2010, central banks are extending their role, using their balance sheets to buy securities,

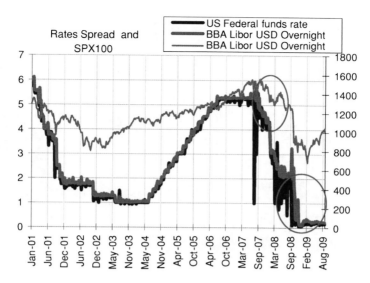

Fig. 4.5 Rates spread (2001–2009). *Source* Bloomberg

monetize government debt and intervene in different markets.[6] Take as an example, the successive interest rate hikes in 2000 by the Federal Reserve to contain the financial bubble, fiscal action that prompted a drop in the equity market. However, fiscal and monetary actions are not always sufficient during an economic slowdown/recession. A decrease in interest rates to stimulate growth is not always sufficient unless it is followed by a decrease in the interbank rates and the rates for businesses and consumers, especially with regards to mortgages. The credit crunch of 2008 is a good example of the inadequacy of lowering interest rates (this goes back to the idea that initial conditions are not similar). Figure 4.5 shows the huge gap between the interest rate established by the Federal Reserve and the interbank rate.

In most cases, the inversion of the rate curve signals a major bear market. This curve is obtained by calculating the difference between yields for long-term government bonds and short- and medium-term bonds. Accordingly, a bearish or bullish trend confirmed by the rate curve leads to the expected decrease or increase of rates and the advent of a recession or expansion. An inversion of the rate curve takes place when the difference between the long-term and the short-term rates becomes negative. Figure 4.6 illustrates the 10 and 2 years spread between 1989 and 2010.

---

[6] For instance, hedging has become a struggle in a world in which so many different outcomes depend on the actions of politicians and central banks. E.g. ban on short selling implemented since the 2008 crisis.

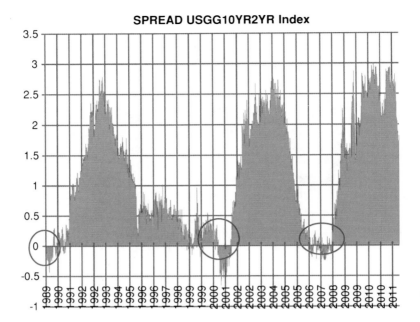

Fig. 4.6 2-year versus 10-year rate curve. *Source* Bloomberg

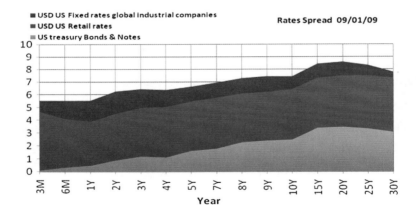

Fig. 4.7 Rate differentials. *Source* Bloomberg

Figure 4.7 illustrates the level of the applicable rates in the U.S. as of 9 January 2009. On this day, the average difference between the rates was 500 bps; markets were frozen and totally illiquid. The question was no longer whether the economy would recover, but whether there was a risk of a severe recession or even a depression.

4.4 Socio-Economic and Other Key Indicators

4.4.1 Society Indicator

Looking for predictive indicator of social unrest, one would look at indicators of economic inequality as is related to the ideas of justice: equality of outcome and opportunities. There are various income inequality metrics (the Gini coefficient, the Theil index, and the Hoover index). Each one has its limits and advantages but when used carefully they can be convenient. That said, a social indicator should not be compiled into one single weighted index but divided by factors that impact economic inequality, as:

- Labor market: unemployment, salaries disparity, jobless youth, income inequality;
- Wealth condensation and taxes policies;
- Education, productive knowledge, and globalization effect;
- Government issues: corruption and lack of democracy, confidence in parliament and population health coverage;
- Working age poverty, elderly and child poverty;
- Population age;
- Burglaries, assault and homicides;
- Life satisfaction and suicides.

4.4.2 Other Key Indicators

Other indicators and ratios are presented in Table 4.2. This list is by no means exhaustive, but rather highlights some of the key indicators.

- Liquidity trap phenomenon: When monetary policy is unable to stimulate an economy, a liquidity trap can occur whereby interest rates decline toward zero but additional cash balances are kept unused. It is a combination of low velocity of money and low interest rates.
- Steel index: Base metals are a good precursor for signaling economic downtrend as they signal the marketplace apprehension. The lead time on base metal prices as an indicator can be short, but there is a pattern that has held up. When using these signals, it is important to watch out for temporarily misleading factors; for instance, prices move due to distortion in supply and demand, trader speculation, monetary easing, or major currency moves which are used as hedges against inflation. It is best to first check what is driving the demand. Also, history shows that they tend to move down long before the economic data show a slowdown. Although this is good signal, it can lead to errors in timing (Fig. 4.8).

Table 4.2 Macro indicators guideline

Global indicators
World trade volume (trade in goods and services) and industrial production
Food and oil prices
World prices (manufacturers)
Consumer prices
Total external debt
M2 to foreign reserves ratio
Debt service
Debt service plus short-term debt as a percent of foreign reserves
Ratio of debt service to recurring revenues
Debt per capita
Short-term debt as a percentage of total debt
Short-term debt as a percentage of foreign reserves
Real GDP growth
Fiscal balance as a percentage of GDP
Current account balance as a percent of GDP
Financial indicators
Government bond spreads (2-year yield spread over German bonds, bps)
Interbank spreads (3-month LIBOR—3-month government bill rate)
Sovereign CDS spreads (1-year bps)
Exchange rates (EUR/USD—EUR/CHF—USD/CHD, USD/JPY)
Equity markets level and rate of increase of asset prices
Volatility
Government bond yields ⇒flight to safety check
Private credit growth bank lending to the private sector as a percentage of GDP
Growth indicators (annualized % change of 3-month moving average over previous 3-month moving average)
Purchasing managers index (PMI—manufacturing % services)
Employment
Estimated change in global inventories
Real private consumption savings as a percentage of GDP
Real gross fixed investment
Investment as a percentage of GDP
Balance sheet and saving rates
Wealth effect: household debt to income ratio and household growth rate of debt stock (annual rate)
Real house price indices
Household net worth (% of gross disposable income); the weak underlying condition of household balance sheets translates their fragility to any minor shock
Non-financial corporations: debt as a share of financial assets

It is complicated to perceive the appropriate interconnectedness of all the aforementioned factors and indicators. It is important to always remind ourselves that the identification of variables, per se, is not enough to forecast a cycle. Rather, the logical chain between these variables and the intensity of their relationships help us predict the occurrence of a cycle. Analyzing the economy sheds light on

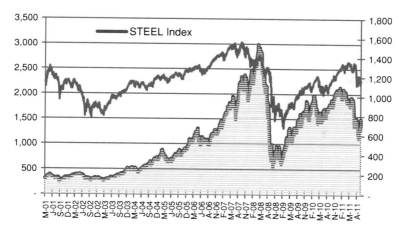

Fig. 4.8 Steel index and SPX index. *Source* Bloomberg

the issue of financial assets allocation. It allows us to draw a forward-looking view that will help us develop risk management strategies, which we discuss in the next chapter.

Chapter 5
Trading Multifractal Markets

Abstract Each market phase has its trading opportunities. A dynamic management approach for trading in multifractal financial markets is introduced in this chapter to allow us to profit from a market's characteristics. An offensive approach is presented based on the notion of diversification at the strategy level between directional and volatility strategies; and of a macro-design approach. Tools such as cyclical and psychological analysis, fundamental convergent analysis, and the estimation of risks, allow us to evaluate the market biases in order to establish an accurate estimation of the prevailing state of the system and the risk toward which it is heading. Once markets' characteristics are grasped, risk forecasting models can be enhanced. Models can be built on the basis of multifractal markets but are not limited to using only fractal tools such as, for example, the Hurst exponent. In fact, fractal thinking allows us to discern the most appropriate way of developing models. Be it technical analysis, behavioral finance, cycle analysis, power laws, thermodynamic, and econophysics, etc....all of these are useful as long as we know how to implement them in our models while remaining aware of their limits. A strategic investment decision must not only be based on the best information available, but also on the possibility of error in the systems of calculation and the development of management strategies. The art of successful tail risk management lies in the ability to hedge against sudden market drifts or any specific micro market risk as well as against long periods of low volatility; and to "time" the volatility to profit from its clustering behavior without having to rely on seismic events to gain profits. That said; total hedging is not possible in absolute terms and if so it does not work at all time. It is important to think in terms of affordable risks before thinking of potential gains. The chapter includes a discussion of recent developments in the various techniques in forecasting risk highlighting their advantages, applications, and limitations.

This chapter examines the concept of portfolio management and risk. It introduces an offensive approach to be ahead of market disarrays and to learn how to thrive in risky markets.

Y. Hayek Kobeissi, *Multifractal Financial Markets*, SpringerBriefs in Finance,   69
DOI: 10.1007/978-1-4614-4490-9_5, © The Author(s) 2013

5.1 Strategies Diversification

Financial instruments and markets have different characteristics and the basis for their classification goes well beyond the volatility measure. Portfolio management and strategy control starts with the definition of a management framework—we proceed with a selection in a context of the optimization of returns, taking into account the investment objectives and risk/gain aversion/seeking profiles.[1] Once we have identified the portfolio management type, we can define investment constraints, investment time frames, and risk tolerance in order to develop a suitable overall investment strategy. The distribution of assets between various forms of revenue including interest, dividends, and capital gains combined with the diverse categories of instruments including money markets, certificates of deposit, bonds, stocks, derivatives, hedges, and leverage, define the degree of general autonomy of the management.

All assets are examined under a rigorous qualitative and quantitative[2] due diligence process to ensure they meet the risk return objectives as part of the overall portfolio design. At this level, overestimation of the advantages of asset diversification by relying for instance only on the beta and historical price series correlations must be avoided. A fund manager uses his experience and research resources to implement a process-driven approach to build efficient portfolios. Proficient portfolios have a robust, dynamic process in place based on strategy diversification and a macro-design approach. Diversification at the strategy level is imperative and is composed of two main elements: directional strategies and market speed strategies.

5.1.1 Directional Strategies

Directional market strategy involves the choice of assets, regions, sectors, and the direction of the position such as long (bull) or short (bear). The specific risk of an asset cannot be reduced. Diversification consists of establishing a portfolio where total losses can be minimized. The beta and Hurst analyses of the portfolio can be useful to a certain extent to determine whether the noise in the portfolio is reduced to validate our diversification. Appendix C presents the RHO as an alternative and as a predictive measure.

That said, asset diversification is completely worthless when markets collide; market dislocation; and liquidity drain causing autocorrelation between asset classes. In this sense, asset allocation as a diversification tool for offsetting portfolio risks is not valid all the time; it generally works only during positive

---

[1] The standard portfolio management types are: conservative, balanced, growth-oriented/dynamic, and aggressive.

[2] Appendix B Equities valuation: FCV.

Table 5.1 Portfolios with different allocations for six asset classes

Portfolios /Allocation of assets	Money market	Bonds	Shares	Real Estate	Commodity (%)	Private equity (%)
1	21 %	35 %	20 %	10 %	14	0
2	6 %	40 %	20 %	14 %	15	5
3	0 %	20 %	35 %	19 %	16	10
4	0 %	0 %	30 %	30 %	23	17
Portfolios	1	2	3	4		
2008 Performance	−18 %	−26 %	−41 %	−52 %		
Historical volatility (1994–2007)	5.83 %	7.33 %	10.09 %	12.26 %		
Historical performance (1994–2007)	7.81 %	8.66 %	9.73 %	10.70 %		

The historical risk return outcomes (1994–2007) of several benchmark asset allocations per portfolio were wrongly assumed as indicators of future performance. Allocations per portfolio were reproduced at the beginning of 2008 and subsequently led to massive losses.

economic cycles; but it is worthless when the directional strategy does not outperform its benchmark and so it cannot justify the management fees incurred. It is important to understand that historical volatility as a measure of risk is misleading, as illustrated by the example in Table 5.1.

A better way to allocate long only strategies is but not limited to adjusting the historical volatility according to the financial cycle period ahead. It is essential also not to forget that simulations of performances are usually based on indices as benchmarks. These indices are continuously attuned and are not representative of factual managers' allocations.

5.1.2 Market Speed Strategies

Market speed strategies involve the trade of volatility (long or short); and the use of financial leverage subject to the derivatives instruments at play. It is important here to know how to trade in a volatile environment while understanding and limiting the risks. Some strategies such as market momentum trading, market neutral trading, hedging, scalping (trading intraday) and event-driven trading can be rewarding if implemented properly. But trying to time the market and the intensity of financial agents sentiments poses huge limits. Morien (2007) emphasizes the limits of market timing stating:

> ...in trading, there is no such thing as too high or too low. There are no overbought or oversold markets. Value, in the Ben Graham context, means nothing to someone that trades a squiggly line. A long term investor knew all along that the Internet bubble wouldn't last and the whole market was quite obviously overpriced in 1987, 1929 and the end of the 60s. Hundreds of traders went broke shorting stocks in the middle of those

booms, trying to pick the moment of reversal. It just didn't work. Jesse Livermore himself was bankrupted for the last time when he went short stocks a few months before the crash of 1929.

The best tool against risks and losses is to operate with appropriate selection methods based on unbiased and pertinent information to identify attractive, undervalued assets, and strategies. In doing so, we limit risk while leaving plenty of room for gains and profits. The key is to know how to achieve superior returns while protecting ourselves from catastrophic events. To do that, we must add another layer of wisdom: Portfolios are safer when built on macroeconomic designs based on cyclical and secular studies, financial behavior patterns, and forward looking catastrophe risk assumptions.

5.2 Tail Risk Hedging: Valuing Risks and Investing in Hedges

In fractals terms, the entire system represents the internal complexity on a larger scale. Thinking in this way helps us to realize that we do not have to hedge each and every position in the portfolio to protect ourselves, but rather, to hedge the entire portfolio and add to it some tactical hedges at the micro level. Consequently, studying and understanding the macroeconomic environment and identifying market cycle phases to hedge against extreme risks can help us avoid huge losses. On the other hand, microanalysis remains vital for generating and protecting short- and medium-term performance from risk. Bhansali (2007) describes this analytical strategy as follows:

> This correlation allows risks to be broadly hedged at the portfolio level, rather than at the security level, which would be cost prohibitive. Think about it this way: When you purchase homeowners' insurance, you don't insure each item of jewellery and clothing and every appliance individually. You insure the entire house and all its contents. Trying to precisely hedge specific risks would be too complicated and too expensive.

In order to evaluate the impending macro risks, we start by identifying the current environment and by establishing forward-looking scenarios. The assessment of risk is to be done the same way we value assets. By doing so, hedges are assessed as an investment. We are not throwing capital out the window; rather, we are investing for when an extreme unknown event occurs. Macro tail risk investment, depending on the size of the position, can offset some or all the losses of the other investment positions in the portfolio.

Adopting these hedges is like having a contrarian macro position with a specific objective, while other strategies and hedges at the micro level are totally independent. The rationale behind the concept of a tail hedge is that probability values do not matter because major events can happen regardless of their probabilities. Rather than relying on the probabilities, we have to prepare for their occurrence by adjusting the hedging proportion in the portfolio. We have to recognize the limitations of our quantitative and qualitative tools and avoid being overconfident.

A well-experienced manager is humble enough to know that combining his qualitative views with quantitative results obtained from mathematical tools of risk-factor analysis, expected volatilities, and correlations is the best way of thinking. By acknowledging the limits of the above-mentioned sensitivity calculation, a manager intuitively fine-tunes his set of tools.

So what are we looking for when assessing risk?

Our objective is to establish a general methodology for estimating risks of extreme sets. Extremes manifest themselves in many ways in finance:

- Extreme losses
- Extreme price swings
- Extreme risk measures
- Extreme correlations
- Extreme diversification breakdown
- Extreme overspills
- Extreme systemic risk

Risk analysis can be divided in three complementary dimensions: the study of sensitivities; the calculation of conditional value at risk (CVaR); and the estimation of "cross–stress" testing scenarios.

Regarding sensitivities factors, it is imperative to take into account the weakest probabilities in order to estimate risk. The source of risk must be identified and decomposed by degree according to the following components:

- The specific risks of an asset, including liquidity risk and credit risk;
- Exchange risk;
- Interest rate risk; and
- Other market risks including contagion risk, liquidity drain, inflation or deflation, and regulatory changes.

Within these categories, risk factors can be identified. Risk factors are elements whose existence and movement cannot be estimated accurately but can nevertheless influence the market price of any given position. In order to estimate risk quantitatively, we can use recurring, observable, and quantitative risk factors which include, for instance, the rate of exchange of currencies, price of a barrel of oil, and level of an equity benchmark index. These factors are based on available information and the choice of risk factors[3] reflects how each one understands the financial market.

---

[3] The sensitivity of a position to the variation of a risk factor measures the power of the risk factor. Here, we list some capital asset pricing model (CAPM) derivatives whose relevance is very limited and should be cautiously used when assessing risk. The beta (β) is the sensitivity coefficient of a stock with respect to its reference index. In mathematical terms, β is the slope of the curve that results from the linear regression between price and market yield, for a given period of time: $dV = \alpha + \beta I + \varepsilon$, $\beta = dV/dI$, where ε is the specific risk and β the systemic risk. As for bonds, duration measures the sensitivity of a bond's market price to yield variations. Convexity, on the other hand, is a measure of the sensitivity of the duration of a bond to changes in yield. We have to assess as well the sensitivity of options in relation to various factors: Delta measures an option's sensitivity to changes in the price of the underlying asset;

5.2.1 The Conditional Value at Risk (CvaR) and the Extreme Value Theory

Value at risk is a static measure that quantifies the potential loss in the value of a risky asset or portfolio based on the hypothesis of the normal distribution of price yield over a certain period and within confidence intervals. What is missing here is a measure that accounts for the severity of failure (the potential size of the loss) and not only the chance of failure. The conditional tail expectation (CVaR) is the expected size of a loss that exceeds VaR. It is determined as follows: Let $X > 0$ be a loss random variable with distribution function $FX(x)$ or density function $fX(x)$. The VaR and CVaR at confidence level p are defined as:

$$VaRp = \inf\{x; \ FX(x) \geq p\}$$

$$CVaRp = E\{X \mid X > VaRp\}$$

CVaR adjusts the VaR to express the expected loss when extreme outcomes occur. It is the VaR plus the expected loss in excess of the VaR. For instance, the CVaR for a 10 % VaR would be the average of the worst 10 % of outcomes.

Extreme value theory[i] (EVT) is a tool for calculating the CVaR. This tool marks a significant step forward in VaR estimation as it makes it possible to evaluate the risks related to shocks that may never have occurred in the past but are more likely to occur in the future. EVT focuses on the modeling of the tails of a distribution that can then be used to estimate the VaR. In fact, EVT deals with the frequency and magnitude of very low probability events. It is the standard tool for hydrologists, reliability engineers, and reinsurers. For instance, reinsurers set premiums in order to cover not only the insured but also to ensure the solvency of the company. According to Smith (2009), EVT (specifically multivariate EVT) applies when we are interested in the joint distribution of extremes from several random variables as it concentrates on low dimensional probability of elements such as winds and waves on an onshore structure; meteorological variables including temperature and precipitation; air pollution variables including ozone and sulfur dioxide; finance variables such as price changes in several stocks or indices; and spatial extremes such as joint distributions of extreme precipitation at several locations, Internet traffic, structural reliability, or biotech analyses.

EVT is fine when dealing with phenomena that we cannot explain very reliably, such as for many natural hazards. But it is not enough for social/financial phenomena, where we have insight into who the participants are in such systems and the effects of their actions (Operational Risks—lack of controls; perverse

---

(Footnote 3 continued)
Gamma measures the Delta's sensitivity to changes in the price of the underlying asset; Vega measures an option's sensitivity to changes in the volatility of the underlying asset; Theta measures an option's sensitivity to time decay; and Rho measures an option's sensitivity to changes in the risk-free interest rate.

incentives; Market risk—strategic moves to control market; central bank interventions) and we can construct, although imperfect, theories to explain these actions (economics, politics, etc.). That said, EVT provides many useful insights on extreme market outcomes, but it is a tool to be used with great care as it is based on extrapolations into the unknown. As Mandelbrot (2005) states:

> It still does not take into account the tendency of bad news to come in flocks or long-term price dependence and crisis coming in quick succession.

A lot depends on judgment and experience. Throughout communication/ correspondences with Professors Neil and Fenton[4] (Martin Neil 2011) we acknowledged that even new models, like Entropy pooling techniques[ii] (where risk managers try to express subjective view on the joint portfolio distribution), Copula Marginal Algorithms (copula is about how to generate joint default scenarios) and Non parametric quantile regression, fails to properly account for the:

- Cluster effect—copula VaR and EVT models do this but only reactively.
- Dependence—the techniques explicitly deal with covariance in many guides but only up to a point, they only go skin deep.
- Amplified feedbacks—to handle this one would need a network model operating at a higher level than EVT and copulas (such as a counterparty default escalation model), however this would be explicitly causal.

Neither of these approaches attempts to tell us why a model has such or such tail behavior nor does it tell the circumstances under which we might observe it. Therefore, we cannot identify early warnings to help mitigate or avoid the tail—it is instead simply a mechanism to price it into our portfolio, according to our risk appetite. Professors Neil and Fenton claim that:

> EVT (especially the peaks over threshold stuff) fails to meet the above requirement because it is purely phenomenological (Martin Neil 2011).

5.2.2 Risk Taxonomy and Risk Cascade Stress Tests

To try to anticipate fat tails, we do not need to produce a long list of events to calculate their likelihood and interrelations. Rather, we can apply a stress test with a short list of the most likely catastrophic events and think about how to prepare for them and what early indicators should we track. Think of this method as playing macro and offense all the time. In other words, instead of wasting time searching for the x percent probability that an event will occur, we can use the following simple but effective techniques:

---

[4] Martin Neil Professor of Computer Science and Statistics and Norman Fenton Professor of Risk and Information management, School of Electronic Engineering and Computer Science, Queen Mary University of London. Their book "Risk Assessment and Decision Analysis with Bayesian Networks", 2012, Taylor and Francis Group is much more of a "how–to" guide using existing Bayesian technology.

- Identify what we need to watch out for. Prioritize and quantify risks and think through the value chain of risks. We achieve this by ordering the risk events: first there are macro events that can affect the portfolio directly (liquidity drain, recession); second there are events that have an indirect impact on the portfolio (credit shortage); and third are the events that have a regulation impact (short-selling ban).
- Improve risk assessment capabilities throughout the economic cycle and not only during bad times so as to better manage risk.
- Think about not only what may happen but also what to do when it happens.
- Plan strategically how to hedge and how to turn some market risks into opportunities.

The advantage of the stress testing technique is that it does not rely on hypotheses; rather, it simulates potential or historical extreme events. Its aim is to determine the outcomes if such extreme scenarios did occur. The scenarios are established as a function of the manager's intuition and judgment and are not a quantitative function of risk. There are no probabilities associated with the scenarios. Stress testing scenarios can be done to assess various types of risk including credit, interest rate, exchange rate, liquidity, and contagion risk. The use of an exponentially weighted stress test is often recommended as it allows us to consider the relative importance of the most similar periods instead of having each period equally weighted. To perform cross-stress tests, we proceed vertically in time, considering factor correlations, and horizontally considering the events of flocks. There are three main types of scenarios that can facilitate risk comparisons across assets:

- Extreme events, which assume the portfolio's return given the recurrence of a historical event. Here we identify a past cycle with reference to the assumed cycle ahead. Then current positions and risk exposures are combined with the historical adjusted factor returns;
- Risk factor shocks. Here the factor exposures remain unchanged, while the covariance matrix is used to adjust the factor returns based on their correlation with the shock factor;
- Events shock indices, (e.g. macro-economic series such as oil prices, or custom series such as exchange rates): by using regression analysis, new factor returns are estimated as a result of the shocks; and correlations between risk exposures (such as, the correlation between a 150 bps move in interest rates and a steepening yield curve).

A stress test is useful because it allows us to qualitatively determine correlation forecasts.[5] Mind that any statistical technique is only an abstraction of reality. Reality cannot be modeled; education, experience, judgment, and intuition cannot be given. The implementation of any model in practice is as much art as it is science where a lot depends on the appreciation of each model's strengths and limitations.

---

[5] The Financial Services Authority (FSA) introduced the reverse stress test on December 2008. An underlying aim of this test requirement is to ensure that a firm could survive long enough after risks have formed either to restructure a business, or to transfer a business.

The mean or variance of a risk measure certain aspects of the risk but do not provide much information about the extreme risk. Value-at-risk (VAR) and volatility are used as a marketing tool to sell a portfolio, fund, or other structures because they provide assurance and simplicity to investors who need to know their risk. Market professionals have used these tools misguidedly. Although, the crises of the twenty-first century have led to changes in risk engineering and tools. It is uncertain, however, whether these changes have helped efficiently predict risks. A more accurate measure of risk must factor in the possibility of loss and its potential magnitude. For transparency and better accuracy on the risk return tradeoffs, funds should reveal in their factsheets and reporting to investors the following data: Downside volatility and Sortino ratio, maximum drawdown, downside correlation, worst and best 12 months, graph of comparative performance during down months, the risk return graph per investment of the fund as well as main peer group and benchmarks etc.

5.2.3 Tail Risk Management

It is costly to carry sizeable hedges during non-catastrophic periods and conversely, not to be hedged when we are in the middle of a market storm. By selecting short to medium term hedges based on short-term instabilities and utilizing long-term hedges based on macro forecasts, we can minimize the effects of bad timing in the portfolio allocation. Nevertheless, we must be careful not to over expose the portfolio to this type of insurance, most especially in periods when it is not required. For instance, when the volatility curve is in contango (upward slope) and since the trades on the volatility are usually done on a monthly basis using futures contracts, we would end up buying high and selling low. Ideally, we should be pro-active in our management by creating vehicles based on either the volatility of the volatility or the shape of the volatility forward curve. The clue here is to find good quality algorithms. Tail risk management is all about adopting active and systematic hedging strategies. Aside from short-maturity treasuries and other cash instruments, long volatility strategies are an effective approach. In tail risk management, volatility can be used as an asset in the portfolio composition. The strategies are outlines as follows:

- Establish a permanent, *defensive, long macro volatility* exposure in line with the portfolio risk budge*t*.
  Choose the appropriate medium-term structure combination (for instance, 3, 12, and 18 months) according to the volatility pattern and actual behavior while taking into account volatility characteristics such as clustering and mean reversion to ensure the cost efficiency of this approach. Like a sailor who stands ready to trim his sails as the wind changes, we must always be ready to increase or decrease our level of protection according to prevailing market conditions in

order to reduce our carrying costs. Our approach to the selection of options should be systematic to avoid any bias.

- Adopt an *offensive volatility strategy* by timing short-term (less than 3 months) market behavior.

The art of successful hedging lies in the ability to hedge against sudden market drifts or any specific micro market risk as well as against long periods of low volatility; timing the volatility and profiting from its behavior without having to rely on seismic events to gain profits. In an extreme market downturn, the correlation between asset classes increases and deepens. In order to cut costs, we can use direct and indirect hedges. The key is to select the appropriate combination of instruments and asset classes. Trade instruments used as hedging tools include: over-the-counter (OTC) variance swaps; call options on low-yielding currencies; exchange-traded products; out-of-the-money options; inflation or deflation structures varying in duration from 10 days to 30 years; and managed futures. We can implement derivative strategies using index options, foreign currency options, credit default derivatives, and derivatives on commodity indexes. However, we need to remember that the safest approach is to stick to the most liquid instrument, which means the most actively traded products; otherwise the hedges would be worthless when needed. It is also recommended to use equity products for hedging credit positions when credit hedges are too expensive.

Based on our brief review of the process of the mind in decision making, we expect that investors will not easily be convinced when asked to invest in protection against a potential catastrophe; that is why, discounting the hedges from income or excess returns, is an approach that may be better received by investors and can be achieved by assigning an account for future hedge strategies when markets are favorable and excess returns are being generated. When excess returns are not available, hedges must be implemented as any other asset class; should we be unable to do so cost effectively, we can buy protection from a fund that sells protection strategies as an asset class. Marketing tools would be effective in convincing investors and creating awareness and acceptance in the same way that most of us pay for medical, house, or car insurance—we do so every year and the only thoughts we have are about the selection of the best insurance provider, the level of insurance and whether or not we want to reduce our annual costs. We should apply the same way of thinking to finance. In the beginning, it was not easy for the insurance industry to convince people to buy insurance until a catastrophe occurred and they realized the enormous costs to which they were exposed.

Investments Banks and asset managers are currently rushing to create complex, volatility-based vehicles. Taleb (2010) argues that these type of tail funds

...Will fall like flies ... we've seen it before[6]

While Bhansali (2009), states

---

[6] See Harrington, Weiss and Bhaktavatsalam (2010) for more details.

The question shouldn't be whether you can afford to hedge, but whether you can afford not to...[7]

The fractal way of thinking encourages tail risk management approaches and gets us to a new level of awareness. Tail risk managers who failed before simply did not implement the appropriate strategies.

5.3 Systematic Control and Offensive Approach

A close collaboration between portfolio managers and their risk unit is vital to be able to profit from market opportunities as well as avoid biases and losses. Portfolio management starts with risk managers implementing clear procedures and a systematic approach. The control management, much like insurance risk-based systems, should:

- Develop two risk interfaces, one at the strategic level and the other at the detailed level and think in terms of discrete changes attaining critical levels. In a sense, the devil is always in the details, so watch out not to override your strategic model because of ever changing details.
- Define normal and catastrophe risk measurement tools that empower decision-making;
- Build catastrophe models based on detection of systematic indicators (i.e. consensus indicators of critical levels) while having a thought through idea of the strengths and weaknesses of those models;
- Adopt the appropriate risk models for each market cycle, for example, Cycle I (high volatility + inflation pressure + sovereign debts pressure) is clearly different from Cycle II (high volatility + inflation pressure + market bubble). Do not apply risk systems where they do not work. Every underlying and every strategy is unique and each requires its own risk approach and has its own specific limits;
- Discretionary management along with a systematic approach is also needed to look for what does not fit and for contradictory evidence. Discretionary intervention is often required when changes in market patterns cannot be detected in time by the algorithm at play; this is when judgment and intuition become vital. For instance, implementation of a Bayesian approach[8] is important to assess market shifts in the weeks ahead. Take as example periods when the release of

---

[7] See Jones (2009) for more details.

[8] According to Smith (2009), to compute the probability of a specific event, a predictive distribution may be much more meaningful than a posterior or likelihood-based interval for some parameter. Bayesian methods are used as a device for taking account of model uncertainty in extreme risk calculations. Only a Bayesian approach adequately provides an operational solution to the problem of calculating predictive distributions rather than inference for unknown parameters in the presence of unknown parameters (Smith 1998).

US unemployment data is expected in the week ahead: the weight given to the unemployment factor in the algorithm should be lower than the normal value when for instance the market attention is on Italian bond markets where yields are exploding beyond 7 %.

To this end, it is essential to establish management standards and to separate the control function from the management one. Still it is crucial that the manager comprehends and prepares for all risks. He is ultimately the only decision maker getting in and out of the market and as such, should be aware of his risks. Much like a chess player moving a pawn requires the reckoning of the reason behind it and the risk in making this move?

When setting limits, we need to specify that which we want to limit and why. That said we need to account for compensation risk between opposite positions, as is the case of straddles and strangles for instance. The risk compensation is determined as Risk of position (1 and 2) < Risk of position (1) + Risk of position (2). However, the latter do not apply to divergent and convergent strategies or to opposite positions with different calendar time horizons.

Once the risk budget is set, a systematic control system is necessary in order to detect possible divergences as soon as they occur. Also, risk-wise we do not have a straight number to control; the best approach would therefore be to look for a risk margin as per the fuzzy logic criteria. Ultimately, the safest way is to implement the correct strategy and adjust it over time, whether systematically or discretionary (intuition is fundamental). We cannot allow ourselves to be rigid in an ever-changing environment. One may use the systematic tool for trade execution to avoid any bias at that precise moment. Discretionary adjustments are crucial for a continuous system update and override when needed according to the pre-set limits. We cannot have a program running through different cycles unless we can feed it continuously (Colyer 2011). This is a mistake often committed by most commodity trading advisors (CTA) managers . After the gains amassed in 2008, they were so over-confident that they did not adjust for changes in the economic and financial cycles and consequently lost a lot in 2009 and 2011. CTA managers need market divergence. In 2011, they bet on catastrophes ranging from Japanese earthquakes, oil reaching USD 200 because of the so-called Arab Spring to potential sovereign defaults in Europe. None of these have yet fully come to pass and so their performance is on the negative side except for some as listed in the below table (Table 5.2).

The solution to management problems in the financial markets is not a magic formula; rather, it is to possess a very good knowledge of the bifurcations of the markets, their butterfly effects and intra-sensitivities. Deterministic and unpredictable laws regulate the financial system. Determinism permits the identification of opportunities by the majority of investors in the period t < B by applying simple directional management methods appropriately. Meanwhile, unpredictability ensures the survival of markets and allows smart managers to identify opportunities. The noisy chaos theory allows us to understand the complexity of the financial system and evaluate the uncertainty that prevails at any phase of the market. The complexity and risks vary from time to time; if the cautious manager

Table 5.2 CTA/managed futures fund performance

CTA/managed futures TRR YR	2011	2010	2009	2008
Winton futures fund LTD-B$	6.3	14.5	−4.6	21.0
Welton-global directional	−14.4	18.7	−5.6	25.3
Diversified trading co II	−6.3	12.7	−7.9	30.8
Aspect diversified fund-USD	4.5	15.4	−11.2	25.4
Transtrend fund-omni US-$	−11.6	20.3	−12.4	32.8
Man ahl diversified futures	−9.2	11.6	−16.4	24.9
Tulip trend fund LTD-C USD	−20.2	38.4	−25.5	60.0

Diagram 5.1 Dynamic management—solutions to multifractal markets using tail risk management

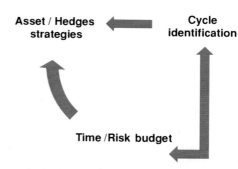

An investor view risk in relation to time.
Time horizon is critical in evaluating the risk of any investment.

can analyze them, he can verify if he can afford them. Survival in multifractal markets rests on the management of risk. It is important to think in terms of affordable risks before thinking of potential gains. The management approach relies on the experience of the manager, his intuition in regard to the most appropriate management plan; his experience and knowledge regarding the market in which he invests and his ability to develop accurate forecasts and to prevent and correct errors in perception. Successful investing requires a close study of the market in order to grasp the opportunities that it presents and avoid losses. It requires us to adopt an offensive approach at all times. Diagram 5.1 summarizes the concepts we have discussed so far and how they apply to our overall strategy.

New thinking and offensive opportunities

"Markets cannot be predicted in absolute terms; it is much easier to predict volatility because it clusters and shows a degree of persistence $(0.7 < H < 0.8)$ (Carbone et al 2004)." While discussing the volatility behavior with Peters (2011), he explained that from a pure Hurst standpoint it is antipersistent. Like the Cauchy distribution, it has no mean or variance. Also it reverts more often than a random process, but it is not "mean reverting" in that there is no mean. However, volatility does have regimes. There are long periods of high and low volatility. They are tied to the business cycle. Davidsson discusses in a research paper (Davidsson 2011) serial dependence from both a theoretical and an empirical perspective. Daily and monthly SP-500 data from 2003 to 2009 are investigated and he finds that returns tend to be serially independent, whereas volatility and expected returns tend to be positively serially correlated. Other findings indicate that daily volatility and expected return data seem to be more positively serially correlated than monthly data. He concludes that:

> ...Any process can deviate from randomness in two different ways: either through drift (expected returns) or through serial dependency (serial correlation). This means that even if returns are perfectly serially independent, we can still find price trends. Most risk in financial markets comes from stochastic changes in expected returns and volatility. Expected return and volatility are not perfectly stochastic, neither perfectly positively serially correlated. The truth lies somewhere in between.

With this understanding, we can identify many opportunities and also avoid many losses. Events and our reactions to them create highly attractive investment opportunities. There are many more arbitrage opportunities than we think. To take advantage of these opportunities, we have to learn how to be on the offensive and exploit market inefficiencies. To limit risk and be on the offensive to exploit it, we need to arm ourselves with multiple layers of defense and then shift to the offensive.

The shortcomings of asset allocation strategies lead us to consider better-performing alternatives such as a macro hedge strategy. The use of this strategy can minimize losses depending on the investment size. Carrying out allocation strategies is as crucial as the choice of assets and underlying instruments. The mix of diverse strategies resembles hedges. Our objective is to benefit from the market cycles and limit the impacts of turmoil from the allocation between hedges, assets, and strategies. The key in achieving optimal combination of strategies is to know how and when to adjust forecasts and positions according to the cycle. The risk of reacting to market noise would be minimized if we change our forecasts quickly enough; otherwise, any position adjustment would be too late. It is important to find the balance between being dynamic and being passive when it comes to market noise.

Appendix B: Equities Valuation (Hayek 2010)

The valuation of equities is articulated around two key elements: the financial soundness of the company and its market price. Our aim in this section is to identify stocks with the greatest promise in terms of revenue, earnings, cash flow as well as positive market predisposition.[9] To do so, we study the financial strength of the company and rate its credit risk profile.

Financial Strength and Credit Rating

The credit rating process involves quantitative and qualitative analysis of a company's balance sheet, operating performance, and business profile. This information is gathered from official company data and meetings with the company management team. There are a number of ratios to choose from for assessing credit risk and it is up to the analyst to identify the more pertinent ratios for a company, its operations, and the industry in which it competes. It is important to look into the company's overall standing, including the company's financial past and present situation to estimate its future outlook. Some of the most widely used ratios for assessing company's strength are: the Earnings per share, the Flow ration, the cash burn rate...etc. These ratios assess how a company uses its available cash, which provides insight into the life expectancy of firms. However, other methods such as evaluating the credibility of the management, measuring the number of projects in development, and examining the quality of the technological platforms are more appropriate. It is also essential to pay adequate attention to the results of products in their first and second phases of development including their efficiency testing.

Once the company's financial solidity is assessed, its fundamental value is compared it to its market value, in order to measure the value gap.

The Fundamental Convergent Price

The fundamental convergent price (FCV) introduced here is the equity adjusted value to the market opinion. It is a forward looking valuation, thereby the notion of convergence of the fundamental value (FV) to the market price and vice versa.

The standard formula[10] for equity valuation assumes that the following factors are known with certainty: the sale price of the share in n years; future dividends;

---

[9] In 2011, for instance, factors such as the aging population and the ever-changing opportunities and innovations in health care technology influence the identification of promising health care stocks.

[10] The price of price of equity is a function of the actualized forecast revenues and the yields corresponding to the investor's investment horizon

$$V_0 = \sum_{i=1} \frac{D_i}{(1+t)^i} + \frac{V_n}{(1+t)^n}$$

and exact discount rates. In reality, however, there is always a difference between FV and the market price, because of market disequilibrium.

The valuation of a company on its own would be useless if we do not add to it its market premium or discount as it is through the market that we execute our trades. Our goal is to calculate a value according to our own valuation while discerning the market's perception. Consequently, most of the distortions and obstructions to the establishment of this value can be eliminated so that the FCV can be estimated. As Keynes (1936) states:

> It is not a case of choosing those [faces] that, to the best of one's judgment, are really the prettiest, nor even those that average opinion genuinely thinks the prettiest. We have reached the third degree where we devote our intelligences to anticipating what average opinion expects the average opinion to be. And there are some, I believe, who practice the fourth, fifth and higher degrees. (p. 140)

People price shares not based on what they think their fundamental value is, but rather on what they think everyone else thinks their value or predicted average value is. The FCV is calculated as:

$$FCV = FV + MPD$$

where FV is adjusted and evaluated according to our own judgment; and

MPD is the evaluation of the market premium or discount, based on our estimates concerning market judgment and forecasts.

The FCV takes into account the phenomenon of convergence and the distortion between the fundamental and market values. At one point in time, we may face two scenarios: either the price approaches the value or the value is modified until it justifies the price. Any of these scenarios can be self-validating. Tvede (2002) describes this process as follows:

> If a stock rises, the company's credit rating improves, as well as its opportunities to finance activities with loans or by issuing new stock... Price fluctuations thus have impact on true value... (p. 7).

The Fundamental Value

In assessing FV, it is necessary to determine a margin of fluctuation for the price, depending on the discount rate, valuation methods used, and growth rate of the forecasted price. The discount rate is the minimum acceptable rate of return or

---

(Footnote 10 continued)

where V_0 is the value of the share on the starting date;

D_i is dividend to be received in year i, with i varying from 1 to n;

V_n is the expected value of the share in year n; and t is the rate of discount.

An investor decides to buy a stock based on whether the market price is less than or equal to V_0.

"hurdle" rate. It is composed of two elements: the risk-free rate and risk premium. These elements reflect both the objective elements which are proportional to the volatility of the share, subjective elements which depend on the sector, quality of the company's management, and the company's competitive advantage. As it is impossible to forecast future dividends from one year to another in the long term; we introduce a second method by calculating V_0 as follows:

$$V_0 = NP \times PER(sector)$$

where NP refers to net profit; and PER (sector) refers to the price earnings ratio of the sector.

Accordingly, V_0 reflects FV of the equity based on the company's net profit realized or forecast according to the rate of growth (g) and the market reference measure, PER, for the sector. We use PER in this calculation because it is the measure most widely used by the market, to minimize the difference between calculated and market price. In deciding which valuation method to apply to forecast dividends, we need to look into analysts' forecasts. Nonetheless, even if all analysts produce the same forecasts, different fundamental valuations for a company can result. This difficulty in arriving at the appropriate valuation method justifies and reinforces the existence of a spread between FV and market price of the equity.

Market Premium/Discount and Limits of Technical Analysis

The market premium or discount (MPD) calculation is influenced by:

- The analysis of economic and financial cycles, where we assess the so-called inner forces which alternatively inspire and discourage human beings. These forces cause intense emotions including fear, anxiety, and greed.
- The technical analysis indicators, in which we examine the behavior that drives decision-making and consequently determines the variations in prices. According to Mandelbrot (2008) it is important to pay attention to the importance given to these indicators, they are a means to an end:

> The short term has essentially been given up to the black arts of the traders, who most of the time make bets on the bets that others are making. Investors love to find patterns and statistical mirages where none exist. (p. 4).

Technical analysis is valuable when studied using the following approach: if part of one's knowledge, hopes, forecasts, revenues, anxieties, and doubts is integrated with the price, then the price chart of a stock also contains part of the existing information with respect to this stock. The efficiency of this hypothesis is its ability to recognize a trend, reveal the characteristics of this trend, and determine its persistency or reversal. Technical analysis is not only another way to interpret the market; rather, it also sometimes influences market psychology and behavior. In reality, market trends are self-fulfilling in the sense that:

...Even if a pattern is entirely coincidental, it may start generating valid signals if enough dealers use it (Tvede 2002, p. 49)

Nonetheless, it is important to be aware that this type of analysis is auto-destructive. Technical analysts recognise long-term dependency where events of the past are reflected in today's prices. However, we need to remember that initial condition changes lead to different results. We cannot copy-paste events and outcomes based on historical behaviors. The key is to not become radical but rather to try to perceive all other indicators. Our mind can play tricks and we end up imagining patterns where there are none. In essence, technical analysts can be right sometimes when trying to guess crowd behavior. We can use their analysis as a tool to trace trade opportunities established by other indicators.

Once the FCV is defined, we can compare it to the market price. The choice of method for avoiding the forecasting trap rests on the choice of financial ratios. In the short- and medium-term, it is necessary to concentrate on known or predictable ratios, compare them to the peer industrial group, or to situate them historically in order to detect a relative under or over valuation[11].

There is no one correct method for market and stock analysis. The breakdown of the price into objective and subjective elements and the instability of the economic and financial systems make the calculation of the market value difficult.

---

[11] The adjusted price earnings ratio (aPER): The most common method for measuring the difference between the fundamental and market values is to calculate the fundamental PER and to compare it to the equity market PER. Another method involves comparing the equity market PER to its peer group. In using either of these methods, note that the ratios are influenced by:(1) the phase of the economic cycle: e.g. it is completely normal that the PER increases in a growth environment; (2) the quality of management, position of the company in its industry and its long-term potential; (3) interest rates; and (4)Speculation or rumors surrounding the company affect its price premium

The self-financing ratio: The PER for some sectors including the media, pharmaceutical and technology sectors cannot be assessed accurately because companies in these sectors invest heavily in research and development and have significant financial needs. As such, it is preferable to base our analysis on the self-financing ratio

Dividends: Companies who get most of their yields from dividends are more sensitive to a change in interest rates than others. In order to keep attracting investors, these companies, well-established and mature, offer high dividends in order to compete with certificates of deposit. This method, however, is criticized because it can be sometimes difficult to identify a specific sector of activity for a given company. The peer group is not always easy to establish and if improperly identified, can lead to false market conclusions. Also, difficulties can arise given the weakness of all statistical and historical approaches and the fact that ratios are static. Take for example the crash of 1990. In *How Can You Tell a Bear Market Is Over* by Birinyini Associates, several indicators including the PER ratio, dividend yields and short and long term treasury bonds, were analyzed in order to identify those which could have detected the floor for the S&P 500 crash in October 1990. The three-month treasury bonds signaled to purchase on 10 May 1991. The long-term treasury bonds, gave the same signal in March 1993. Neither the PER nor the dividend yields gave such signals regarding the end of the bear market in early 1990.

We must be conscious of the difficulty of this exercise and approach the FCV by situating it in a margin and proceed with forecasts to regularly monitor and manage the portfolio.

Given the forecasting dilemma, it is better to forego long-term forecasts and to use short- and medium-term forecasts in estimating the FCV. The gap between equity market price and fundamental value creates an uncertainty in the financial environment and makes all portfolios vulnerable to risks factors. Bear in mind, that while it is important to find out if a share is correctly valued, it is critical to know how this evaluation will be modified over time with the movement of the economic and financial cycles (the strange attractors). In this sense, Bernd Engelmann & Daniel Porath article (Engelmann & Porath 2012) "Do not Forget the Economy when Estimating Default Probabilities" is quiet interesting as it introduces techniques to integrate macroeconomic information into a rating model and then illustrates how the macroeconomic variables improve the performance of a model for small- and medium-sized companies.

> Rating systems without macroeconomic information run the risk of yielding imprecise probability default (PD) estimations....Modular approaches, like e.g. the Bayes approach generally are more flexible. In principle the Bayes approach can be applied to any rating system, even to pure expert ratings and incorporate forward looking components.

Appendix C: In the Lab

Researches and most intriguing in the lab findings that we came across are presented hereunder.

Predicting Risk/Return Performance Using Upper Partial Moment/Lower Partial Moment Metrics (2011d) (Viole & Nawrocki, Predicting Risk/Return Performance Using UPM/LPM Metrics (2011d).

In their paper, Viole and Nawrocki develop a better explanatory/predictive measure that takes lower as well as higher moments into account. Below is a summary of their rationale.

- Below target analysis alone is akin to only hiring a defensive coordinator.
- Diversification is the panacea of risk management techniques to reduce the non systemic risk relative to an individual position. The preferred method to reduce nonsystemic risk is to add investments with the greatest historical marginal non systemic risk net of systemic risk. This has been quantified by subtracting the systemic benchmark from the investment as in the below equation.

$$\text{Non Systemic} \qquad\qquad \text{Systemic}$$

$$\left(\frac{UPM(q,l,x)}{LPM(n,h,x)}\right) \quad - \quad \left(\frac{UPM(q,l,y)}{LPM(n,h,y)}\right)$$

- This ratio answers the question when comparing and ranking multiple investments simultaneously: What investment historically goes up more than the market when the market goes up and historically loses less when the market loses—And should I even be invested in this asset class?
- Systemic risk. Autocorrelation/dependence/serial correlation ($p(x)$) is an important tool in identifying increasing distributional risks such as muted entropic environments and lending a predictive ability to an explanatory metric. The autocorrelation formula for a 1 period lag is for investment x at time t is:

$$|\rho(x)| = |cov(xt, xt-1)|$$

- The absolute value is used because an autocorrelation of -1 or 1 is equally dangerous to investors. A 1 period lag is used because we aim to err on the side of caution. Where there's smoke there's fire. If a 10 period lag presents autocorrelation, it will obviously be noticed in the one period prior. The risk is that between lag differences, a bifurcation abruptly ceases thus leaving the investor waiting for a confirmation to avoid the very event that has just transpired, effectively rendering this metric explanatory.

Explanatory   Predictive

$\left(\dfrac{UPM(q,l,x)}{LPM(n,h,x)}\right)$ $\quad - \quad |\rho(x)|$ $\left(\dfrac{UPM(q,l,x)}{LPM(n,h,x)}\right)$

- An observed one autocorrelation reading denotes a dubious situation. The increased autocorrelation influence can be subtracted from itself to compensate for an increased likelihood of an unstable investment, thus lowering the metric to reflect this probable risk. The trick is to know when to exit prior to the bubble $\rho(x)$ is our predictive metric using the autocorrelation coefficient. It is not intended to pick an inflection point (since one never knows the top until after they have seen it), but the translation to deltas onto your position will properly manage anticipated risks. Then replacing the investment with a senior ranked investment is a viable interpretation of the data.

Our empirical test generates rank correlations between the performance measures and asset returns for both an explanatory period and an out-of-sample predictive period. On balance, we were able to generate minimal explanatory correlations with no out-of-sample predictive correlations using the explanatory model and no explanatory correlations with significant out-of-sample correlations when using the predictive model. Our use of statistical certainty $\rho(x)$ as a punitive variable in quantifying risk humbles the methodology in so far as admitting we are beholden to an uncertain future per Frank Knight. This is an important point as we use certainty as a punitive variable and distinguish between explanatory and predictive functions for the metrics; it has always been just explanatory in nature. Coupled with the degrees in the UPM/LPM measure it is the most behavioral statistic of behavioral finance we have come across.

Do Bayesian Methods Help Model Uncertainty?

As per discussion with Professor Neil, the Bayesian approach uses the prior and likelihood distributions to produce the posterior distribution for parameters of interest. The posterior predictive distribution is the posterior probability of the event/proposition based on the parameters. So, from a purely mechanical perspective it is adaptive to real data and can respond non-monotonically, if necessary, by selecting hypotheses that are better supported by empirical events. These properties are fundamental to "common sense" reasoning i.e. the model can change "its mind". It favors predictions that are most often correct and admits to uncertainty (by virtue of the fact that everything is a distribution of belief).

> For extreme risk calculations I would say that one might need to mix imagination, history, data and insight to produce a causal explanatory model. For this pure Bayesian statistics is not enough (i.e. just using Bayesian parameter updating as an alternative to maximum likelihood won't cut it). That's why there is a lot of interest in using Bayesian networks to model the causality structure that might better predict extremes and norms, as mixture models of market epochs/states. Data then helps identify which hypothetical state/epoch the market is in and the beauty of the Bayesian approach is that this data can take any form (expert opinion, market data, harbingers etc.).
>
> Doing this is, computationally hard, especially if you are aiming at an alternative to CAPM.

Any model should strive to be explanatory as well as prognostic, hence our stress on causal attribution and the explicit role of systematic causal factors in the models. What we really need is a way of building into our models discrete measures ("in control", "out of control" etc.) to track, monitor, and signal predicted risks. The research work of Professors Neil and Fenton fuses causal modeling, copulas and portfolio aggregation into a single Bayesian framework—early results are promising, but there is a still lot to be done. Portfolio rebalancing starts by setting up the proper portfolio control tools and periodic portfolio follow-up procedures at the outset.

Scale of Market Quakes in High Frequency Data

Dupuis and Olsen (Dupuis and Olsen 2011) propose a different way to analyze high frequency data: an approach in which the time series is dissected based on market events where the direction of the trend changes from up to down or vice versa. Physical time is replaced by intrinsic time (ticking at every occurrence of a directional change of price) where any occurrence of a directional change represents a new intrinsic time unit. The scale of market quakes (SMQ) defines a tick-by-tick metric to quantify market evolution on a continuous basis.

In the probability density function $\mathrm{Log}(f(t)) = I.\delta.t - \gamma.|t|^{\alpha}$ $(1 + I.\beta.(1/|t|).\tan(\alpha.\pi/2)$

For $1 < \alpha < 2$, δ (local parameter) is defined as the mean, and the variance is infinite.

Estimators which involve only 1st powers of the stable variable have finite expectation, like the fractile ranges and absolute mean deviation, are appropriate measures of variability for these distributions than the variance.

In Olsen Ltd. model, the fractile ranges are estimated as the intrinsic timeλ, and the absolute mean deviation ϖ (δ_i) is the quantile averages of the scale magnitude quake (SMQ). Since α and β remain constant under addition, the means δ_i and the scale parameters γ_i can be identified (12 scaling laws[12] have been discovered by Olsen Ltd.)

> ...The scaling laws are powerful tools for model building: they are a frame of reference to relate different values to each other... The scale of market quakes is an objective measure of the impact of political and economic events in foreign exchange and used as a support tool for decision makers and commentators in financial markets or as an input for an economic model measuring the impact of fundamental economic events. The scale of market quakes can be used in different ways; decision makers can use the indicator as a tool to filter the significance of market events. The output of the SMQ can be used as an input to forecasting or trading models to identify regime shifts and change the input factors.

Their bet is that imbalances will correct themselves... this is where the model finds its limit: what happens when imbalances does not correct or takes more time to do it?

Market Behavior Near the "Switching Points"

Preis and Stanley (2011) examine whether concepts from physics to discover if there are general laws to describe market behavior near the "switching points" in the data. To do so they analyzed massive data sets (transactions recorded every 10 ms of the German DAX Future stock market, daily closing prices of stocks in the S&P 500 share index in the US) comprising three fluctuating quantities—the price of each transaction; the volume; and the time between each transaction and the next—to find out if there are regularities either just before or just after a switching point.[13] Under the Panic hypothesis, their analysis revealed that the volume of each transaction increases dramatically as the end of a trend is reached, while the time interval between each transaction drops:

> In other words, as prices start to rise or fall, stock is sold more frequently and in larger chunks. Traders become tense and panic because they are scared of missing a trend switch.

---

[12] This is the equivalent of a Richter scale in geology for financial markets. The list of scaling laws is on page 11 of Dupuis & Olsen paper.

[13] These are either local minima where the share price falls before starting to rise again (also known as an "uptrend") or local maxima where the price peaks before falling (a "downtrend").

Cascade Dynamics of Price Volatility

Alexander M. Petersen, Fengzhong Wang, Shlomo Havlin, and H. Eugene Stanley studied the cascade dynamics of price Volatility (Petersen, Wang, Havlin, and Stanley., 2010) immediately before and immediately after 219 market shocks.[14] The results of their paper are of potential interest for traders modeling derivatives (option pricing and volatility trading) on short time scales around expected market shocks, e.g., earnings reports.

We define the time of a market shock Tc to be the time for which the market volatility $V(Tc)$ has a peak that exceeds a predetermined threshold. The cascade of high volatility "aftershocks" triggered by the "main shock" is quantitatively similar to earthquakes and solar flares, which have been described by three empirical laws—the Omori law, the productivity law, and the Bath law. We find quantitative relations between the main shock magnitude $M = \log_{10} V(Tc)$ and the parameters quantifying the decay of volatility aftershocks as well as the volatility preshocks....Information that could be used in hedging, since we observe a crossover in the cascade dynamics for M 0.5. Knowledge of the Omori response dynamics provides a time window over which aftershocks can be expected. Similarly, the productivity law provides a more quantitative value for the number of aftershocks to expect. Finally, the Bath law provides conditional expectation of the largest aftershock and even the largest preshock, given the size of the main shock. Of particular importance, from the inequality of the productivity law scaling exponents and the pdf scaling exponent for price volatility, we find that the role of small fluctuations is larger than the role of extremely large fluctuations in accounting for the prevalence of aftershocks.

---

[14] We analyze the most traded 531 stocks in U.S. markets during the 2 year period of 2001–2002 at the 1 min time resolution.

Chapter 6
The Latest "Normal"

Abstract Today, we are in the midst of an undesirable mix of tight economic and financial conditions. Opaque derivative products which include among others Collateralized-Debt-Obligations and Exchange-Traded Funds continue to shape the markets. Economists need to rethink certain economic concepts and relationships. Actually, fiscal stimulus spending is not stimulating anything. The deadly embrace between over-indebted sovereigns and over-leveraged banks has created a vicious cycle. Faced with these new economic and financial market imbalances, portfolio management has become much more difficult. We need to arm ourselves with how to manage tail risks against the inevitable and recurring loss of confidence which comes from market instability. We must adapt continuously along with the market whatever the pressure. Acknowledging errors and adjusting to them is crucial. Understanding market events and their effects on the market system is an essential element toward building a financial warning system. To this end, we need to identify all pre-switching points; implement them in algorithms that can trigger a warning signal. If we think of Keynes levels, the sixth level would be to have a drummer like mind ... to be ahead of the beat. In the end what we achieve by understanding markets' fractality is more than a tool; it is a way of thinking.

The financial system constantly changes and for each period there is an accepted "normal" market. As such, we need to pay attention to new market characteristics, often initially identified as irregularities, because they have the ability to substantially change the market structure and function. Changes to the "normal" market do not just happen out of the blue, they are the product of an accumulation of historical events and memories that either brings about a gradual change to the perceived normal market or as in the recent past, the changes are brought about by seismic events that abruptly and significantly change the market.

The market we live in now has come about as a result of the financial crisis that began in 2007 and which largely resulted from the inability of regulators, central banks, and large market participants to fully understand the complexities that had been introduced into the market on the back of innovations in trading methods and

Y. Hayek Kobeissi, *Multifractal Financial Markets*, SpringerBriefs in Finance, DOI: 10.1007/978-1-4614-4490-9_6, © The Author(s) 2013

complex derivative products whose introduction was driven by the need for market participants to shift risk and cleanse their balance sheets in order to meet regulatory liquidity requirements.

6.1 Black Boxes

Economies around the world are more vulnerable to events and shocks, including butterfly effects, which are widespread and have different intensities at different times. The increase in volatility is accentuated by the access to the markets of all kinds of investors and financial agents, including those that are inexperienced. The transition from a market dominated by long-term investors to one dominated by short-term investors has greatly modified the concept of portfolio management.[1] Opaque derivative products which include Collateralized-Debt-Obligations (CDOs) and Exchange-Traded Funds (ETFs) continue to shape the markets.

6.1.1 CDOs: Collateralized-Debt-Obligations

Innovations in computer technology facilitated the development of mathematical modeling tools that have created complex financial products, especially opaque derivative products which include:

- structured products from the simplest "plain vanilla" structures to the most complex products which allow real client costs to be pocketed by issuing banks and hidden from clients; and,
- products issued from the securitization of assets with little liquidity including CDOs. Banks developed tactics for freeing up their debts by "securitizing" them or offloading them to investors in the form of CDOs and their derivatives including CDO squared, CDO cubed and synthetics, in order to tighten their balance sheets and abide by their regulatory ratios. CDOs are debt titles issued by an ad hoc structure called a "securitization vehicle", which buys and retains bonds issued by companies or banks. Credit default swaps (CDS) were created to guarantee the buyer against the risk of default.[2] Where, the buyer pays a premium periodically and the premium is expressed in annual or nominal percentage in the case of CDS spreads.[3] The seller thus commits to indemnify the buyer of the CDS in case of a credit event affecting the underlying entity

---

[1] Brokerage firms encourage this type of aggressive management because they significantly increase the volume of daily transactions and consequently, their fees.

[2] This is similar to the purchase of a put except that CDS are bilateral contracts.

[3] The amount of the premiums (or spreads) paid to a given issuer gives an indication of the market appreciation and the quality of this issuer.

such as default or restructuring as stipulated in the contract. These contracts are exchanged by mutual agreement in opaque and unregulated markets. A fundamental aspect of CDS is therefore the compensation. Appendix D explains the events of the 2008 meltdown.

6.1.2 Exchange-Traded Funds

The surge in ETFs, exchange-traded notes (ETNs) ,and related products was a sign of flow into the passive industry management. Unable to outperform the markets, index trackers are packaged and sold to the public with the bonus of daily liquidity. But what are exactly these products?

A 2009 ETF conference in Amsterdam proved to be informative and included interesting participants with conflicting opinions on important matters such as counterparty risk and settlements and regulations which left the audience confused and troubled. ETFs are essentially passive funds, designed to track the movements of an index. Some ETFs are physically-backed, that is, they hold the underlying investment which they are supposed to be tracking, whereas the majority of ETFs use synthetic (swap-based) replication to facilitate their exposure. ETFs have many inherent risks and as they become more complex and blurry, the intrinsic risks become less obvious. In addition to the market risks that any ETF investor is exposed to, there are other inherent structural risks in this exponentially spreading instrument:

• Counterparty risk
  In the case of swap-based ETFs, the ETF provider enters into swap contracts with other financial institutions rather than obtaining direct exposure to the target index. Under UCITS III regulations, ETFs have to be at least 90 % covered by collateral; theoretically this means that there is only a 10 per cent counterparty risk with another provider. Nevertheless, the ETF can fail if the ETF trades cause a huge loss in a counterparty that does not have sufficient capital to cover that loss and honor the derivative contract. On top of that, the swap contracts between a synthetic ETF provider and an investment bank are constructed in a "black box", making it virtually impossible to understand how the target returns are delivered.
  But, are physically-backed ETFs safer? They might have been if the majority of the providers did not practice securities lending, the process of lending securities to a third party in exchange for a fee. The type and quality of collateral requested from borrowers in these transactions can and does vary to the extent that the fund can end up facing major liquidity issues during a financial crisis. Securities lending can result in huge losses occurring if the borrower defaults. According to Bioy (2011),

  ... in a sense, securities lending can simply be seen as making physical replication an approximate mirror opposite of synthetic replication: you start with a perfect basket and you turn it into an imperfect basket by accepting collateral in exchange basket and use a swap to receive the performance of a perfect basket.

- Tracking error risk

  Providing extra protection to investors frequently results in additional costs. This is reflected in the performance of the ETF in the form of a negative tracking difference between the return of the underlying index and that of the ETF. Many, if not most, of the ETFs optimize their basket to replicate the benchmark index they are supposed to be tracking. As such, these ETFs obviously display greater tracking error risk when volatility spikes. ETF performance does not, therefore, always match the underlying index. As a result of daily rebalancing and compounding, the leveraged and short ETFs can sometimes fail to adequately track their index and can miss sharp rallies or can aggravate falls in the corresponding index.

- Spread gap risk

  The liquidity of ETFs is a serious and separate issue from the liquidity of the underlying shares. The trading spread gap can vary between 10 and 50 bps depending on the chosen ETF provider. Considering that investors rush into this type of instrument in order to trade a specific index instead of investing in a mutual fund, most investors find themselves negatively surprised at the cost they incur when executing their trades.

- Settlement risk

  The UBS scandal in September 2011 which reported losses of $2.3 billion is an ideal example of this type of risk. The UBS trader involved in September 2011 scandal was capable of establishing fake transactions in UBS's system to disguise his exposure largely because of the weak trade settlement rules in existence in the London market and the fact that many counterparties do not automatically request trade confirmations. In addition, the bank's risk control-lers failed to check the huge trading positions that were reported[4] as hedges and their subsequent margin calls. According to Amery (2011), Europe's ETF market participants need to make sure of the following:

  All bilateral, off-exchange trades in ETFs must be reported so that the average investor can get a fair idea of what's going on. Clearing houses and settlement systems should publish data on the efficiency of the post-trade processes in ETFs just as their counterparts do in the US market and in the same way as Europe's stock exchanges regularly publish data on bid-offer spreads and turnover. Such data should be published both for individual funds and for ETF market makers, so that we can see where potential problems lie. Without such steps, public confidence in the ETF market, which must already be at low ebb after the UBS scandal, is likely to wane further. John Bogle, founder of Vanguard and the father of index-based investing, repeated his long-standing criticisms of ETFs in a CNBC interview ..., calling them 'a bastardised version of the index fund' and adding that 'only an idiot would want to trade indices all day in real time. The onus is on the ETF industry to prove him wrong.

---

[4] iShares began recently in 2011 to publish a detailed quarterly report outlining all its ETFs' securities lending activities.

Despite their risks, ETFs continue to attract substantial asset inflows from other areas of the financial market because they are a relatively easy-to-use instrument for gaining exposure to a variety of different markets. According to Morningstar fund flows, at the end of September 2011, there were $970 billion of assets under management in U.S. ETFs, compared to just $904 billion the year before. Some argue that ETFs have enabled investors to access a wide range of asset classes that were previously limited to sophisticated investors. It has helped drive down costs in the passive investment industry, however, while cost went down, risks went up.

Some instruments or structures should only be available to the professionals who are qualified and experienced in running the required proper due diligence. Even if more information is disclosed in the future, the risks will never be sufficiently understood by the general retail market. It is very possible that in the near future ETFs will create the same banking meltdown as CDOs did in 2007, unless more targeted regulation is applied.[5] Many of the potential risks that have been high-lighted here are not specific to ETF's—they concern other managed fund structures as well. In the case of the ETF's, the problem is amplified because they are accessible to uninformed and non-professional investors. How can we expect them to perform appropriate due diligence? More regulation is needed and for once and for all banks should take liability for their employees' attitude. In the food pro-duction industry, a firm cannot escape and hide behind an employee who happened to fail to check, for example, the perished milk that was being used in the production of a chocolate bar and then simply apologize for any resulting health problems on the basis that one of their employees made a mistake. Strict rules are applied inside the firm and outside of it to ensure that this type of mistake does not happen. Risk control systems that keep on failing in the banking system make me wonder if the banks are really failing or if top management simply ignores discrepancies in the greedy pursuit of higher returns. Ethics are not an abundant characteristic.

6.2 The Battle of the Economists

The aim of this section is not to compare the thoughts of famous economists', rather it is to attempt to pinpoint and include their thoughts and apprehensions about the latest economic normal. We will leave the discussions and battles to skilled economists to articulate.

Accepting the new socio-economic and political environment, economic norms and global power equilibrium involves an analysis of the indebtedness of devel-oped countries versus the solvency of the emerging ones. Nowadays, the market is focused on the possibility of the default of several top rated governments in Europe, Japan, and even the U.S.

---

[5] Especially european regulations. In the USA, the US 1940 Investment Act, require an ETF to own the constituent assets of the index it is tracking to be classified as a fund.

As a result of political risks and a rising debt burden Standard & Poor's lowered their U.S. long-term rating to "AA+". Similarly, sovereign debt strains in Europe have lowered the credit ratings of Italy, France, Greece, Spain, Ireland, and Portugal. In February 2012, Moody's announced it had placed the UK rating on a negative outlook largely due to slow GDP growth. Meanwhile, as the profiles of other European countries continued to worsen, and with the entire Eurozone under threat of dilution, the global banking sector is also facing the challenges of new regulations and increased capital requirements that will result in the need for new capital injections. Inflation, leading to stricter capital controls in emerging markets and potential stresses in the various bond markets, is threatening yet more uncertainty and the real possibility of a global economic downturn. The increased likelihood of persistent stagnation in developed economies is likely to lead to socio-political unrest as already evidenced in Greece.

During the first half of 2011, there were many catalysts for economic growth, for example, the easing of the U.S. Federal Reserve rates and Federal government spending. However, these Federal policies targeting assets did not help matters. Deflation continues to threaten governments that are trying to work their way out of debt. In 1933, Fisher addressed the causes and effects of depressions in his paper entitled "The Debt-Deflation Theory of Great-Depressions" and he also discussed possible solutions, in summary he stated:

> ...The dollar may swell faster than the number of dollars owed shrinks; the more the debtors pay the more they owe... In that case, liquidation does not really liquidate but actually aggravates the debts and the depression grows worse instead of better. By March, 1933, liquidation had reduced the debts about 20 %, but had increased the dollar about 75 %, so that the real debt, that is the debt as measured in terms of commodities, was increased about 40 % [(100 − 20 %) × (100 + 75 %) = 140 %]... Finally, I would emphasize the important corollary of the debt-deflation theory, that great depressions are curable and preventable through reflation and stabilization. (Fisher 1933)

Although reflation is difficult to implement in practice usually because actual indebtedness is colossal, the vicious cycle of deflation and a failing credit system cannot be stopped if deflation is not controlled, reversed and the credit system restarted as explained by Fisher. As of February 2012, the benchmark interest rate in the United States was at 0.25 %. Since the 2008 crisis, the US government has provided liquidity to the markets but it could not incentivize banks to lend again and thereby to restore confidence in the system. Incentives had been largely ignored in fear of impending high capital reserve regulations. Regulation should urgently address these concerns in order to get the economic wheels turning and not itself become part of the cause for persistent stagnation. This should be done while at the same time avoiding implementation of controls that would further constrain the functioning of the financial system.

For instance, the decision to keep Federal Reserve rates low until mid 2014: what will this decision achieve? Nothing palpable in the near term except that interest rate certainty will help to finance the banking system. A suggestion would be to also impose low interest rates on loans attributed to productive knowledge and sustainable investments, in other words fiscal and monetary stimulus has to be selective in its objectives, in order not to fall in the Hayekian rebuke that stimulus only stimulates deficit. As Hayek (Hayek 1994) succinctly noted:

> The more the 'state' plans; the more difficult planning becomes for the individual (p79)

In developed markets, central banks are currently offering virtually-free money and companies are investing relatively little—they are hoarding their cash. The fact is that individuals cannot be left in a pure "laissez-faire" system.[6] Therefore, governments should provide an institutional framework within which decisions are left to individuals on what to do and how to earn a living. Regulation is nevertheless necessary for the market to function and for individuals to carry out their plans. Governments cannot, however, plan the details as they have no knowledge about the future, let alone about the present.[7] In his famous book "Capitalism and Freedom", Friedman (1962, 2002 edition 1) states:

> The Great Depression, like most other periods of severe unemployment, was produced by government mismanagement rather than by any inherent instability of the private economy.

We can assume that in an ideal world all that governments would plan for would be pure "Biased Policies". Consequently, their role would be complimentary—each country according to its own structure—by implementing regulations with the aim of:

- preventing the advent of negative externalities.[8] This would include ensuring market transparency within the framework of anti-fraud regulations, not limited to investment firms and extended to the practices of rating agencies for instance;
- providing controlled leveragecontrolled leverage[9]: leverage being the only control variable as clearly explained by Dr. Woody Brock in American Gridlock

---

[6] Same principle applies in religion: religious laws have to be taken as general guiding rules but should not ordinarily intrude in an individuals' daily life. Once a society reaches the level of detailed governmental guidance, it is on its way to servitude and can then be easily manipulated into extremism.

[7] Questions persist regarding governments' ability to control financial activities given the difficulties of evaluating and accounting for complex derivative products or illiquid assets.

[8] Some activities cannot be regulated by the market place as they can cause damage to third parties or negative economic outcomes known as negative externalities.

[9] Jerome L. Stein's paper on Stochastic Optimal Control (SOC) (Stein, 2010) Analysis focused on the question: What is an optimal debt or leverage that maximizes the expected growth of net worth and is based upon significant risk aversion? The Stochastic Optimal Control Analysis provides another tool of analysis and derives the time varying optimal debt ratio.

1. The optimum debt ratio or leverage maximizes the expected growth of net worth.
2. As the debt ratio rises above the optimum, the expected growth of net worth declines and the risk rises.
3. The probability of a crisis is positively related to excess debt, equal to the difference between the actual and optimal debt ratio, measured in standard deviations.
4. An unambiguous early warning signal (EWS) of a debt crisis would be that the leverage $f(t) = L(t)/X(t)$ exceeds f-max, so that the expected growth of net worth is negative and the risk is high.

(Brock, 2012). Conditions that lead to a perfect storm or maximum endogenous risk are correlated forecast mistakes, pricing models uncertainty, problematic hedging, incomplete markets, and leverage will always be there even when greed and self-dealing play no role. All of the variables with the exception of "Excess Leverage" are state variables.(p 129)

- providing elasticity in bank reserves parameters, this can be achieved if banks' and funds' reserves become dependent on flexible margins set by economic and financial conditions such as an increase of reserves in the occurrence of economic reheating and vice versa;
- accommodating the wager between exemptions and fiscal surcharges, where fiscal exemptions can be efficient in encouraging consumption and investment if applied distinctly to key sectors; otherwise they simply lead to an increase in public debt and to social unrest;
- promoting socially-responsible investments and productive knowledge.
- Enabling positive changes from re-channeling liquidity to crucial economic sectors. Ideally, liquidity should be accompanied by a rise in employment in order to fully realize economic benefits; and
- creating safety nets to avoid social unrest.

...and always hoping that a War will not be the chosen solution.

The difficulty with the above would be to know who should follow-up on the implementation of regulations and be responsible for disclosing any discrepancies as warnings to the public. Handing this role solely to government raises a conflict of interest, because when central banks lack independence (political dependence), they tend among other foolishness to inflate the money supply.

Adequate control systems at the governments and at the corporate levels are needed to improve the socio-economic world we live in... at least to avoid social unrests.

The financial crisis of the past decade should encourage us to rethink some economic concepts and relationships. Today's consumer is very different from his counterpart in the 1970s. We are all credit consumers, whether we use credit to purchase household goods, transportation, or education. We often overspend to sustain our lifestyles, using up our lines of credit when, in reality, we do not have the means to pay for our purchases. The everyday investor has become focused on gains in order to sustain his consumption. A change in econometric formulations is thus imperative; economists must rewrite certain relationships or else we will end up incorrectly reading into many market indicators. In 1993, Shiller proposed to rely more on indexation to reduce the impact of inflation or deflation on the economic GDP share and recently in 2009 together with Kamstra (2009), they proposed the use of a small denomination GDP share which they named "the Trill":

It would be structured as simple as shares in corporations, with coupon payments instead of dividends that would rise in an expansion and would decline in a recession with declining tax revenues, in contrast to existing debt vehicles.

According to Stiglitz: new methods for calculating the GDP to be better able to estimate the global society's well-being are needed (Stiglitz 2009, 2010). Neglect of real balance and development has led us to the current status quo. Better measures of growth and the establishment of new statistical factors are crucial.

Faced with these new economic and financial market imbalances, portfolio management has become much more difficult. We need to arm ourselves with how to manage tail risks against the inevitable and recurring loss of confidence which comes from market instability. We must adapt continuously along with the market whatever the pressure. We should ask ourselves: how does a handicapped person deal with life? How can a deprived population enjoy life? How can we continue to live after the death of beloved ones? Well the answers to those questions are all the same, **we just adapt**. We think that we cannot withstand some situations before they actually occur, but our mind is so powerful; we just concede to the new situation unconsciously and evolve consciously. Otherwise, we would have long ago ceased to exist.

Appendix D: 2007 Meltdown (Hayek 2010)

In the following, we will see how the greed of financial agents, the pure laissez faire of authorities regarding leverage and their inability to understand complex financial operations and the indifference of the masses, combined to bring about the financial meltdown of 2007, the effects of which continued to be felt across the globe at the time of this book going to press.

The bankruptcies seen in 2008 were only a symptom of the difficulties that business banks had been dealing with since 2007. These signals, largely ignored or camouflaged by the financial community[10], were comprised of the following:

- Monetary and leverage laxity;
- A surge in toxic assets which arose from real estate loans given to households with modest incomes;
- unfaithful rating agencies, apparently concerned only with increasing their profitability and consequently losing their objectivity[11]; and
- deficient risk management systems, based on the neoclassical school of thought that proposes easy but ineffective methods of forecasting risks.[12] Specifically in

---

[10] Since the beginning of the mortgage crisis in 2007, Central banks have injected liquidity (through bank loans, at very short terms), in vain.

[11] These agencies have, since 2000, lightened their rating criteria and modified their evaluation models.

[12] This pertains to the greed of portfolio managers and all other control cells (such as the case of Kerviel of the Société Générale bank and many other management platforms who registered heavy losses from the failure of risk follow-up either at the level of internal information procedures or of calculations). It should also be noted that the investments in internal technology systems and platforms in financial institutions have not been sufficient.

the case of CDOs; these were based on Li's Gaussian copula function put forward in early 2000 which assumed that correlation was a single constant number. Rating agencies and banks grabbed the opportunity and shaped their assessments based on this function. While the formula was new and complicated for non-mathematicians, banks were happy to classify their products according to this single risk measure and as such sell their CDO products without much effort. It is impossible for professionals at the time not to have been aware of the danger of this function's limitation as correlation cannot be stable over time.

When the subprime mortgage loan bubble burst, it led to a banking crisis. As of January 2010, defaults on mortgages that had been given to the least credit-worthy borrowers drove financial institutions worldwide to undertake $1.8 trillion in write downs and losses. The following section outlines the significant events during the 2007–2009 market crash.

2007

- 5 March: First signs that the speculative real estate bubble had burst was seen at HSBC, the largest European bank in terms of equity market price, which declared large losses in the U.S. mortgage credit market following repayment defaults.
- 2 April: New Century Financial Corporation declared bankruptcy; they were the second largest U.S. subprime lenders, specializing in providing credit to the most disadvantaged.
- 17 July: Investment bank Bear Stearns Companies, Inc. announced the failure of two of its speculative subprime funds.
- 19 July: The Governor of the U.S. Federal Reserve, Ben S. Bernanke, announced in front of the U.S. Senate that losses arising from subprime products were close to $100 billion.
- 9 August: French bank BNP Paribas announced the freezing of three of its investment funds exposed to the subprime market as the bank was no longer able to evaluate them. The equity markets fell and several days later, the U.S. Federal Reserve lowered key interest rates.
- 14 September: The Bank of England provides an emergency loan to Northern Rock Plc to prevent its bankruptcy.
- October: UBS and Citigroup announced they had been affected by the crisis. American International Group (AIG), the largest global insurer, announced that it was in great difficulty and that it could no longer cover its obligation to guarantee credit defaults. It had sold $441 billion of unsecured coverage without enough reserves.

2008

During the first half of 2008, financial agents speculated on raw materials, penalizing companies and consumers. In the U.S., the increase in interest rates from only 1 % in 2004 to 5.25 % in 2008 and the progressive decrease in real

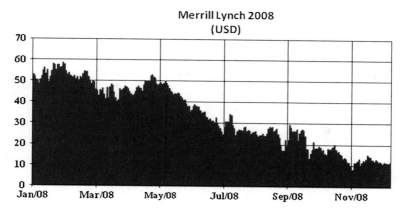

Fig. 6.1 Merrill Lynch stock price falls from $50 to $10 in 2008 (*Source* Bloomberg)

estate prices placed many individuals in difficult times as they were no longer able to repay their loans.[13]

- 16 March: Under pressure from the U.S. government, JP Morgan Chase bought Bear Stearns Companies, Inc.
- 12 August: UBS AG, the largest Swiss bank, experiences massive withdrawals of deposits. The bank then announced its plan to separate its investment entity from its portfolio or wealth management entity.
- July and August: The equity market prices of Freddie Mac and of Fannie Mae, two U.S. quasi-government mortgage refinancing firms, collapsed.
- 7 September: The U.S. Treasury guaranteed the debts of Freddie Mac and Fannie Mae up to $100 billion.
- 15 September: Lehman Brothers, the fourth largest U.S. investment bank, went bankrupt, while Merrill Lynch (Fig. 6.1) was purchased by Bank of America to avoid bankruptcy.
- 16 September: AIG was rescued by the U.S. government who granted it $85 billion in assistance in exchange for 80 % of their capital.

Table 6.1 shows the collapse in equity prices of some financial companies between January 2007 and 2009. Globally, central banks reacted by lowering their key rates and injecting liquidity in an attempt to revive the credit market.

- End of September: Bankruptcy risks spread to Europe. Fortis, Dexia, and l'Hypo Real Estate were among the affected banks.

---

[13] During the real estate boom, debtors refinanced their mortgages under the best of conditions given the over-valuing of their mortgages, or sold at inflated prices. This was no longer possible after the bursting of the real estate bubble.

Table 6.1 Financial institutions' equity market prices

Bank name	Market cap loss, Jan 2007–2009 (%)
Credit suisse	−70
JP Morgan	−50
Goldman sachs	−63
HSBC	−58
Citigroup	−93
Bank of America	−82
AIG	−90
Barclays	−91
Société générale	−74
BNP paribas	−72

Source Bloomberg

- October: The US Federal Reserve accepted to exchange risky securities from banks for liquidity. A U.S. rescue plan named the Troubled Asset Relief Program (TARP) was put into place.

During the course of the rest of the year, the equity markets around the world continued to collapse; Iceland declared bankruptcy; the ruble was devalued; the IMF granted huge loans to Pakistan; Great Britain announced a rescue plan for banks in the amount of GBP 250 billion which was further increased by GBP 100 billion in January 2009. The G-7 committed to help to prevent any bank from going bankrupt, while large EU countries developed national rescue plans worth EUR1.7 trillion to refinance banks[14].

Despite all the help that they received, banks remained cautious with regard to extending credit lines. As such, central banks picked up the slack by lending at longer terms (such as, for instance, up to 6 months instead of the usual three) and accepting non-liquid and badly-valued assets as security. In the end, central banks found themselves with "bad" assets, in a market where in the words of Michel Aglietta:

...capitalism functions as usual, a privatization of profits and a socialization of losses (Michel Aglietta 2008).

Some countries considered the possibility of guaranteeing all interbank loans, while others such as France opted to create a refinancing bank that would lend to the banks on the markets, guaranteed by the state, for long durations between 1 and 5 years, while using wide margins of credit as a collateral. Other countries such as the U.S. repurchased commercial bills from companies for their daily operations in order to fulfill their operating funding needs. These measures, however, did not prevent equity prices from deteriorating and liquidity from staying frozen. In 2008,

---

[14] In the process, however, these nations forgot about other companies in need.

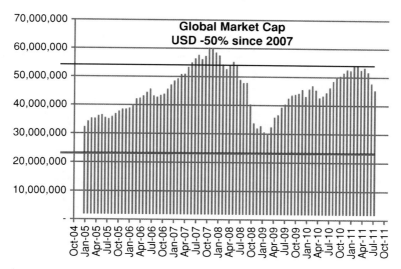

Fig. 6.2 World capitalization (*Source* Bloomberg)

the world equity market capitalization was reduced by 50 % from US$60 to US$30 trillion Fig. 6.2.

Financial companies lost more than US$1 trillion in the credit market following the burst of the subprime bubble (Table 6.2).

Equity markets got into a spiral dominated by the short-term and distorted by the weight of the number of players, specificity of computer programs and margin calls. Figure 6.1 shows how volatility reached new highs during the crisis Fig. 6.3.

2009

Companies with enormous needs for operating capital such as General Motors went bankrupt. Meanwhile, individual consumers' financial situations were also deteriorating, as evidenced by high default rates on credit card payments.

- January 2009: the assimilation of information in prices was high. The market enters a bear phase where economic data was immediately taken into consideration. The system was moving toward its economic attractor. While the economic indicators remained pessimistic and uncertain, the market remained in a bearish consolidation where severe declines were registered and rebounds were confined within a technical margin of consolidation. Trading strategies inside the bear channel were very profitable because the rebounds that followed after several days of decline were often large and vice versa. In short, the gambler's fallacy approach during this time was profitable. Market players wondered whether they were entering into a depression phase and whether they would suffer a fate similar to or worse than that of Japan or the Scandinavian

Table 6.2 Financial losses

Region	Cumulative financial losses June 2007–January 2009 (USD BN)
World	1058.7
Americas	730.3
Europe	297.3
Asia	31.1

Source Bloomberg

Fig. 6.3 SPX volatility index (VIX) (*Source* Bloomberg)

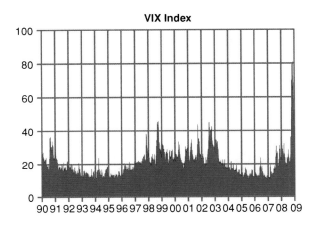

countries in the 1990s.[15] At the beginning of 2009, the hedge fund industry, much like the banking industry, knew it had a lot of work to do to regain investors' confidence, following the large losses of the majority of funds' fraud cases such as the Ponzi scheme of Bernie Madoff and the restrictions imposed to prevent withdrawals by investors in 2008.

• March 2009: Government injections of liquidity reached US$1.6 trillion in September 2009 and signs indicating that a depression had been avoided led to a surge in equity prices. Still, the rate of unemployment in the U.S. and E.U. was 9.5 %, compared to 18.5 % in Spain. In a deflationary environment with USD LIBOR close to 0 %, it was absolutely necessary for the bullish trend to be validated by new market data, lest the market put itself at risk again with another crisis. Indeed, the enormous injections of liquidity had allowed us to avoid a worldwide economic depression, but the economic recovery was slow and without vigor. Until employment figures return to normal, the world economy will remain stagnant and governments vulnerable to social disturbances. In September 2009, the GNP for the U.S., E.U., China, and India were $-3.90, -4.70, 7.90$, and 6.10 %, respectively. During this period, governments had to deal with enormous

---

[15] In 1990, Sweden nationalized several banks and created a "bad bank" which bought toxic assets after discount, leaving the financial institutions to manage their more liquid assets.

budgetary deficits; and with the level of interest rates as low as they were, governments no longer had any tangible means for rescuing markets.

- Between March and October 2009, the S&P500 and EUROSTOXX600 indices increased by 51 and 48 %, respectively, on the back of signs of economic rebounds including the return of consumer confidence and stabilization of real estate prices. The market, however, was still plagued with uncertainty regarding how the market deals with this euphoria; the outlook for the coming months; and how forecasts are affected by the newly-established rules and laws that were exogenous or endogenous to the system. Due to these remaining uncertainties, some governments went so far as to forbid the short-selling of some classes of shares to control speculation which prejudiced funds that use coverage strategies.

Conclusion

Complex interconnectedness in the financial system leads to under-pricing of risk. This status quo will persist as long as human beings exist; it is part of how markets function.

Forecasting is more precise when we analyze the evolution of the markets in a less restrictive manner and if we consider the macro environment rather than just the immediate, micro environment. In practice, opportunities can be found in complex models; hence they exist in the financial markets if we are able to identify the multifractal nature of the financial system and the noisy chaos process that it generates. During this process, local or partial determinism is introduced by the identification of memory effects in the short and medium term. Meanwhile, unpredictability is introduced by recognizing the sensitivity of the system to initial conditions and by identifying the triggers (information that has not been disseminated, correlated investment horizons and leverage levels) of the bifurcation entropy process which evolves in parallel to the Lévy-Stable process. Because market prices change at irregular time intervals, measurement of market activity in terms of discrete time "t" needs to be adaptive. Calendar time ought to be replaced with intrinsic time according to each specific investment horizon.

Tools such as cyclical and psychological analysis, fundamental convergent analysis, and the estimation of extreme risks, allow us to evaluate the market biases in order to establish an accurate estimation of the prevailing state of the system and the risk toward which it is heading. Since financial markets strive to foresee economic fluctuations, it is crucial to be able to identify the characteristics of the economic cycles and to analyze those specific characteristics. The identification of macro-state factors (economic cycles, disequilibrium, and changes), understanding the market's characteristics, and the accurate evaluation of the sensitivities of a portfolio to these same risk factors are fundamental tools for all investors wanting to guard against the sudden reversals of trends.

Once the markets' characteristics are understood, risk forecasting models can be built and enhanced. Models can be built on the basis of multifractal markets but not limited to specific fractal tools—for instance the Hurst exponent. In fact, the fractal thinking allows us to discern the most appropriate way of developing models. Be it technical analysis, behavioral finance, cycles analysis, power laws, thermodynamic and econophysics, etc., all of these are useful as long as we know how to implement them in our models while remaining aware of their limits.

Experience teaches us how to minimize market risks and manage portfolios efficiently in a rather complex market. A strategic investment decision must not only be based on the best information available, but also on the possibilities of error in the systems of calculation and the development of management strategies. All portfolio management is subject to risks and evaluation errors, either in the identification of the specifics of the economic and financial cycle in question, the level of the calculation of the value of an asset, or the degree of the sensitivity of the portfolio to risk factors. Once all of these measures are in place, we need to adopt offensive strategies that include macro tail risk management strategies to face unexpected market events. To be able to achieve and protect the desired "alpha" in a market, it is imperative to only take on the risks that can be afforded and to account for the true magnitude of risk. Not all strategies can be hedged and we can never implement a perfect hedge.

Survival in multifractal markets rests on the management of risk. It is important to think in terms of affordable risks before thinking of potential gains. Apart from the fact that we will all be dead in the long run (Keynes 1936), there is another thing that we can be sure of: outperformance will be wiped out due to market dislocations. Risk measures require the modeling of possible variations in market prices according to probable and extreme circumstances. Standard statistical models cannot capture the emotional probabilities or intensities of investors including their greed, fear, and sudden consensus reversal. Learning how to build new risk models, that describe extreme events while taking into account diverse risk factors, is essential. The implementation of any model in practice is as much an art as it is a science; we acknowledge that diversified strategies should not be restricted to asset diversification. To minimize the occurrence of reactionary decisions, we need to diversify our strategy directions (both convergent and divergent), diversify market speed strategies, and implement hedges as an asset in our optimal allocation.

If you are not a high-frequency trader, you must exercise caution and look out for HFT signals. This heightened awareness can be thought as the fourth most important step in the Keynes context[1]. Meanwhile, the fifth step consists of the quick interconnection and assessment of all of the above measures for executing one's strategy or updating one's systematic program if needed. Acknowledging errors and adjusting for them is crucial. Consensus detection in the algorithms and

---

[1] "It is not a case of choosing those (faces) that, to the best of one's judgment, are really the prettiest, nor even those that average opinion genuinely thinks the prettiest. We have reached the third degree where we devote our intelligences to anticipating what average opinion expects the average opinion to be. And there are some, I believe, who practice the fourth, fifth and higher degrees" (Keynes 1936, p. 140).

convergence extrapolation imply that risk warnings are triggered on time. Understanding market events and their effects on the market system is an essential element toward building a financial warning system. To this end, we need to identify all pre-switching points; implement them in algorithms that can trigger a warning signal in advance. The sixth level is to have a drummer-like mind ... always think ahead of the beat.

In the end what we achieve by understanding markets' fractality is more than a tool; it is a way of thinking. This book describes a continuing effort to improve investment management through the understanding of market processes and the adoption of the fractal approach. I hope that it will provide the elements to develop enhanced management tools. Models of investment and risk management can always be improved upon.

References

Carbone, A. G. C. (2004). Analysis of clusters formed by the moving average of a long-range correlated time series. *Physical Review E,69*, 026105.

Alexander, M., Petersen, F. W. (2010). Market dynamics immediately before and after financial shocks: quantifying the omori, productivity, and Bath laws. *Physical Review E, 82, 036114*.

Allais, M. (1953). Le comportement de l'homme rationnel devant le risque: Critique des postulats et axiomes de l'école Américaine. *Econometrica,21*, 503–546.

Amery, P. (2011, September 21). *The ETF loophole (almost) everyone missed*. Retrieved from IndexUniverse.eu.

B, M. (2010). *Fractals and scaling in finance, discontinuity, concentration, risk*. NY: Springer.

Baillie, R. T., Bollerslev, T., & Mikkelsen, H. O. (1996). Fractionally integrated generalized autoregressive conditional heteroskedasticity. *Journal of Econometrics,74*, 3–30.

Barber, B. M., & Odean, T. (2006). All that glitters: The effect of attention and news on the buying-behaviour of individual and institutional investors. *Forthcoming in The Review of Financial Studies*.

Barranger, M. (2000). *Chaos, complexity, and entropy*. Cambridge: Massachusetts Institute of Technology.

Bayes, T. (1763). *An essay towards solving a problem in the doctrine of chances*. Philosophical Transactions of the Royal Society of London 53.

Bayes, T. (1763). *Philosophical transactions of the royal society of London*, Vol. 53.

Belisle, T. (2011, October 4–5). CATCo investment management. (Y. HK, Interviewer) Montreux.

Bhansali, D. V. (2007). Dr. Vineer Bhansali discusses PIMCO's approach to tail risk hedging. *Q&A February*.

Bhansali, D., & Jones, L. M. (2009, April/May). Investors should mind their tail. Undiscovered managers.

Bilger, B. (2011, April 25). The possibilian: What a brush with death taught David Eagleman about the mysteries of time and the brain.

Bioy, H. (2011, September 22). Physical ETFs: A call for transparency. *Hemscott News*.

Black, S. (1976, December). Rational response to shocks in a dynamic model of capital asset prices. *American Economic Review*, pp. 767–779.

Bookstaber, R. (2011, December 12). *Rick bookstaber*. Retrieved from Rick Bookstaber: http://rick.bookstaber.com/2011/12/volatility-paradox.html.

Boulding, K. (1981). *Evolutionary economics*. Beverly Hills: Sage Publications.

Brock, D. H. (2012). Dr. Horace "Woody" Brock American Gridlock: Why the right and left are both wrong—commonsense 101 solutions to the economic crises Wiley, January 11, 2012. NY: Wiley.

Buff, P. (2011). *Investing with volume analysis*. FT Press.

Canada, T. C. (2009, September). *The conference board of Canada.* Retrieved from http://www.conferenceboard.ca/HCP/Details/Society/jobless-youth.aspx?pf=true.

Carbone, A., Castelli, G., & Stanley, H. E. (2004). Analysis of clusters formed by the moving average of a long-range correlated time series. *Physical Review E,69*, 026105.

Chavez-Demoulin, V., & Roehrl, A. (2004, January). *Extreme value theory can save your neck.* Retrieved from Approximity, risk management: http://www.approximity.com/papers/evt_wp.pdf.

Chorafas, D. (1994). *Chaos theory in the financial markets.* Illinois: Irwin professional publishing.

Chorafas, D. (2010). *Risk pricing.* London: Harriman house.

Colyer, R. (2011, October). The trader meets the robotrader. Retrieved from October 11.

Cozzolino J. M., Zahner, M. J. (1973). The maximum-entropy distribution of the future market price of a stock. *Operations Research*, 1200–1211.

Dacunha-Castelle, D. (1996). *Chemins de l'aléatoire, le hasard et le risque dans la société moderne.* France: Flammarion.

Damasio, A. R. (2002, September). Remembering when. *Scientific American,* pp. 66–73.

David Nawrocki, T. V. (2011). *A Bifurcation model of market returns.* Villanova: Villanova University working paper.

Davidsson, M. (2011). Serial dependence and rescaled range analysis. *EuroJournals Issue 64.* International Research Journal of Finance and Economics.

Derman, E. (2011, Octobre 28). Intuition, initial and final. Retrieved from Reuters: blogs.reuters.com.

Derman, E. (2002). The perception of time, risk and return during periods of speculation. *Quantitative Finance,2*, 282–296.

Dinkel J. J., Kochenberger, G. A. (1979). Constrained entropy models: Solvability and sensitivity. *Management Science*, pp. 555–564.

Dormeier, B. (2011). *Investing with volume analysis: Identify, follow and profit from trends*: FT Press.

Dupuis, A., & Olsen, R. (2011, October 18). High frequency finance: Using scaling laws to build trading models, Olsen Ltd. Essex: Centre for Computational Finance and Economic Agents (CCFEA), University of Essex, University of Essex, UK.

Eagleman, D. (2011, April 18). Ask the author live: Burkhard bilger on time and the brain. The New Yorker.

Ekeland, I. (1995). Le Chaos. DOMINOS.

Embretchs P., K. C. (2001). Modeling extremal events for insurance and finance. Springer (VII).

Engelmann, B., & Porath, D. (2012). Do not forget the economy when estimating default probabilities. Retrieved from Wilmott Magazine Article: http://www.wilmott.com/pdfs/120216_engelmann_porath.pdf.

Fama, E. F. (1963). Mandelbrot and the stable paretian hypothesis. *Journal of Business,36*, 420–429.

Fisher, I. (1933). Debt-deflation theory of great depressions. *Econometrica*, 337–357.

Friedman, M. (1962, 2002). Capitalism and freedom. Chicago: University of Chicago, edition 1.

Friedman, M. (2002). *Capitalism and freedom.* Chicago: University of Chicago; (1962, 2002 edition 1).

Fuller, R. J. (2000). Behavioral finance and the sources of alpha. *Journal of Pension Plan Investing 2*(3).

Gls, S. (1952). *Expectation in economics.* Cambridge: Cambridge University Press.

Graham, J. R. (2006). Value destruction and financial reporting decisions. *Financial Analysts Journal,62*(6), 27–39.

Gross, L. (1982). *The art of selling intanglibles: How to make your million ($) by investing other people's money.* New York: New York Institute of Finance.

Groth, J. (1979). Security-relative information market efficiency: Some empirical evidence. *Journal of Financial and Quantitative Analysis,14*(3), 573–593.

Guiasu, S. (1977). *Information theory with applications*. New York: McGraw Hill.

Harrington, S. D., Weiss, M., & Bhaktavatsalam, S. (2010, July 19). Pimco sells black swan protection as wall street markets fear. *Bloomberg*.

Hayek, F. A. (1994). *The road to serfdom*. Chicago: University of Chicago Press.

Hayek, Y. K. (2010). *Marches fractals: Stratégies d'investissements*. Paris: Publibook-Connaissances et Savoirs.

Heiner, R. (1983). The origin of predictable behavior. *American Economic Review,73*(4), 560–595.

Holthausen, D. M. (1981). A risk-return model with risk and return measured as deviations from a target return. *American Economic Review,* 71(1), 182–188.

Jeffrey, R. H. (1984). A new paradigm for portfolio risk. *Journal of Portfolio Management, 11, No. 1, Institutional Investor*, pp. 33–40, Fall.

Kaen, F., & Roseman, R. (1986). Predictable behavior in financial markets: Some evidence in support of heiner's hypothesis. *American Economic Review,76*(1), 212–220.

Kahneman, D., & Tversky, A. (1972). Subjective probability: A judgment of representativeness. *Cognitive Psychology 3*, 430-454.

Kahneman, D., & Trevsky, A. (1974a). Judgement under uncertainty: Heuristics and biases. *Science, 185 & 211 (4481)*, 453–458.

Kahneman, D., & Tversky, A. (1992). Advances in prospect theory: Cumulative representation of uncertainty. *Journal of Risk and Uncertainty,5*, 297–323.

Kahneman, D., & Tversky, A. (2000). *Choices values and frames*. Cambridge: Cambridge University Press.

Kahneman, D., & Tversky, A. (1974). Judgment under uncertainty: Heuristics and biases. *Science,185*(4157), 1124–1131.

Kamstra, M. J. (2009). *The case for trills: Giving the people and their pension funds a stake in the wealth of the nation*. Cowles Foundation Discussion Paper No. 1717, August.

Kant, I. (1787). *Of Space and time*.

Kant, I. (1787). *Critique of pure reason*. London: Bohn's phislosophical library.

Keynes, J. (1936). *General theory of employment interest and money*.

Lawrence Jones, M. A. (2009, April/May). Investors should mind their tail. Undiscovered managers, 140.

Lazlo, E. (1987). *Evolution: The grand synthesis*. USA: New Science Library.

Lee, C. M., & Swaminathan, B. (2000). Price momentum and trading volume. *The Journal of Finance.* 55(5), 2017–2069.

Levy, P. (1937). *Theorie de l'addition des variables aleatoires*. Paris: Gauthier-Villars.

Li, D. X. (2000). *On default correlation: A copula function approach* (pp. 99–107). NY: Riskmetrics working paper number.

Libet, B. (2004). *Mind time, the temporal factor in consciousness*. Cambridge: Harvard University Press.

Liubeckiene, L. (2009). Subjective views and stress-tests for optimization of credit risk of loan portfolio. *Quantitative Finance 325804, Erasmus University Rotterdam*.

Lorenz, E. (1963). Deterministic non periodic flow. *Journal of Atmospheric Sciences, 20*.

Mandelbrot. (2005, May/June). (C. M. Christopher M. Wright, Interviewer) Real estate portfolio.

Mandelbrot. (1997). Fractales, hasard et finance. Flammarion.

Mandelbrot. (1967) How long is the coast of Britain? Statistical self-similarity and fractional dimension. Science **156**, 636–638

Mandelbrot, B. (2008). *The (mis)behaviour of markets—a fractal view of risk, ruin and reward*. London: Profile books.

Mandelbrot, B. (1960). The Pareto-Lévy law and the distribution of income. *International Economic Review,1*(2), 79–106.

Mandelbrot, B. (1963a). The variation of certain speculative prices. *The Journal of Business,36*(4), 394–419.

Mandelbrot, B. (1963b). The variation of certain speculative prices. *Journal of Business,36*, 394–419.

Mandelbrot, B. (1967). The variation of some other speculative prices. *Journal of Business,40*, 393–413.

Mandelbrot, B., & Milhench, C. (2004, November). Claire milhench, fractal finance for the masses. *Global Investor*.

Mandelbrot, B., & Ness, V. (1968). Fractional brownian motion, fractional noises and application. *SIAM Review,10*, 422–437.

Mandelbrot, B., & Taylor, H. W. (1967). On the distribution of stock price differences. *Operations Research,15*, 1057–1062.

Mandelbrot, B., Fisher, A., & Calvet, L. (1997). A multifractal model of asset returns. Cowles Foundation Discussion Yale University Paper #1164.

Martin Neil, N. F. (2011, November 10). EVT limits. (Y. HK, Interviewer).

McNeil, A., Frey, R., & Embrechts, P. (2005). *Quantitative risk management concepts, techniques and tools*. Princeton: Princeton University Press.

Meucci, A. (2011). A new breed of copulas for risk and portfolio management. *Risk,24*(9), 122–126.

Michel Aglietta, B. T. (2008, September 15). It is the failure of a model. *L'express*.

Morien, T. (2007, January). Travis morien's investment FAQ. *The Gambler's Fallacy*.

Morse, D. (1980). Asymmetrical information in securities markets and trading volume. *Journal of Financial and Quantitative Analysis* (December 1980): 1129–1148.

Murphy, R. (1965). *Adaptive processes in economic system*. London: Academic.

Musser, G. (2011, September 15). Time on the brain: How you are always living in the past, and other quirks of perception.

Nawrocki, D. (1984). Entropy, bifurcation and dynamic market disequilibrium. *The Financial Review,19*(2), 266–284.

Nawrocki, D. N., & Harding, W. H. (1986). State-value weighted entropy as a measure of investment risk. *Applied Economics,18*(4), 411–419.

Nawrocki, D. (1995). R/S Analysis and long term dependence in stock market indices. *Managerial Finance,* 21(7), 78–91.

Nawrocki, D., & Vaga, T. (2012). A bifurcation model of market returns. Villanova University working paper.

Neslehova, J., Embrechts, P., & Chavez-Demoulin, V. (2006). Infinite mean models and the LDA. *Journal of Operational Risk,1*(1), 3–25.

Nicolis G., I. P. (1977). *Self-organization in nonequilibrium systems*. New York: Wiley.

Odean, T. (1998, October). Are investors reluctant to realize their losses?. *Journal of Finance* 53(5), 1775-98-1998b.

Paraschiv, C., & Chenavaz, R. (2011, February 25). Seller and buyers reference point dynamics in the housing markets. *Housing Studies*.

Pellionisz, A. J. (2002). *FractoGene IP portfolio*. Retrieved from fractogene: http://www.fractogene.com/.

Peters, E. (1991). A chaotic attractor for the S&P 500. *Financial Analysts Journal*

Peters, E. (1989, July/August). Fractal structure in the capital markets. *Financial Analysts Journal*.

Peters, E. (1992, November/December). R/S Analysis using logarithmic returns: A technical note. *Financial Analysts Journal*.

Peters, E. (1996). *Chaos and order in the capital markets*. New York: Wiley.

Peters, E. (2011, December). Volatility behavior. (Y. H. K, Interviewer).

Petersen, A. M., Wang, F., Havlin, S., & Stanley, H. E. (2010). Market dynamics immediately before and after financial shocks: Quantifying the omori, productivity, and bath laws. *Physical Review E,82*, 036114.

Phelps, E. A. (2004). Human emotion and memory: Interactions of the amygdale and hippocampal complex. *Current Opinion in Neurobiology,14*, 198–202.

Phillips. (1991). *Washington post*

Sauvage, G. (1999). Les marchés financiers; entre hasard et raison: le facteur humain. *SEUIL*.

Schaumburg, J. (2010). *Predicting extreme VaR: Nonparametric quantile regression with refinements from extreme value theory*. Retrieved from SFB 649 Economic Risk: http://sfb649.wiwi.hu-berlin.de/.

Schulz, H. (1981). *Bear market investment strategies*. USA: Dow Jones-Irwin.

Schumpeter, J. A. (1939). *Business cycles. A theoretical, historical and statistical analysis of the capitalist process* (Vol. 461). New York, Toronto, London: McGraw-Hill Book Company. *Scrapbook of reviews.*

Shefrin. (2000). *Beyond greed and fear: Understanding behavioral finance and the psychology of investing*. Oxford: Oxford University Press.

Shefrin, & Statman. (1985). The disposition to sell winners too early and ride losers too long: Thery and evidence. *Journal of Finance* 107.

Shefrin, H. (2011). AIF executive course. Amsterdam.

Shefrin, H. (2011, June). AIF executive course. Amsterdam, Netherland.

Shiller, R. (1993). *Macro markets: Creating institutions for managing society's largest economic risks*. New York: Oxford University Press.

Smith. (1998). Measuring risk with EVT. University of North California.

Smith, R. L. (2009). Notes for STOR 890 course, Chapel Hill, Spring Semester.

Soros, G. (1994). The alchemy of finance. Wiley Investment Classic.

Stanley, T. P. (2011, May). Bubble trouble: Can a law describe bubbles and crashes in financial markets? *Physics World,* 29–32.

Stein, J. L. (2010). *Alan greenspan, the quants and stochastic optimal control*. Retrieved from Economics Discussion Papers, No 2010–2017, Kiel Institute for the World Economy: http://www.economics-ejournal.org/economics/discussionpapers/2010-17.

Stiglitz, J. (2010). *Freefall*. New York: W.W. Norton & Company.

Stiglitz, J. (2009, September 13). The great GDP swindle. *The Guardian*.

Swift Trade, FSA/PN/075/2011 (2011, May 06), UK.

Taleb, N. (2010).

Thaler, R. H. (1999). Mental accounting matters. *Journal of Behavioral Decision Making 12*, 183–206.

Thaler, R. H. (1985).

Thaler, R. (1999b). Mental accounting matters. *Journal of Behavioral Decision Making,12*, 183–206.

Thom, R. (1972.). *Structural stability and morphogenesis*. Benjamin-Addison Wesley.

Thompson, N. (2004, November 7). Fractals, the stock market and our very chancy world. *Los Angeles Times*. Los Angeles, CA, US.

Triple.Net. (2001–2011). Retrieved from cycles in the economy: http://www.triple.net/cycles.php

Tvede, L. (2000). *The psychology of finance*. New York: Wiley.

Tvede, L. (2002). *The psychology of finance*. New York: Wiley.

United Nations. (2007). *Public governance indicators: A literature review*. New York: Department of Economic and Social Affairs ST/ESA/PAD/SER.E/100 United Nations publication.

US Geological Survey. (2011, October 21). *Metal industry indicators composite indexes of leading and coincident indicators of selected metal industries for August and September—Summary Report*. Science for a changing world.

Vaga, D. N. (2012). *A bifurcation model of market returns*. Villanova: Villanova University working paper.

Verrecchia, O. K. (1991). Trading volume and price reactions to public announcements. *Journal of Accounting Research,29*(2), 302–321.

Viole, F. (2011, December). Entropy and fractals. (Y. Hayek, Interviewer).

Viole, F. (2012, Jannuary). Subjective utility and entropy. (Y. Hayek, Interviewer).

Viole, F., & Nawrocki, D. (2011, August 9). *An analysis of heterogeneous utility benchmarks in a zero return environment* (2011b). Retrieved from SSRN: http://ssrn.com/abstract=1907209

Viole, F., & Nawrocki, D. (2012, January 13). *Embracing the cognitive dissonance between expected utility theory and prospect theory* (2011c). Retrieved from SSRN: http://ssrn.com/abstract=1984661

Viole, F., & Nawrocki, D. (2011). Predicting risk/return performance using UPM/LPM metrics (2011d). SSRN http://ssrn.com/abstract=1550519.

Viole, F., & Nawrocki, D. (2011). The utility of wealth in an upper and lower partial moment fabric (2011a). *The Journal of Investing, summer 20*(2), 58–85.

Wright, C. M. (2005, May/June). Mandelbrot interview: Real estate portfolio.

Yudknowsky, E. S. (2003). *An intuitive explanation of bayesian reasoning.*

Zadeh, L. (1965). Fuzzy sets. Information and control, No. 6.

Zaheh, L., & Kacprzyk, J. (Eds.). (1992). *Fuzzy logic for the management of uncertainty.* New York: Wiley.

Index

Endnotes

[i]*Lorenz simple computerized meteorological prototype*

Lorenz constructed a simple computerized meteorological prototype, depending on only 12 factors, in order to carry out numerical simulations. The principle which governed the prototype was very simple: using quantified data from a meteorological photograph, the computer could calculate small projections. The data from these projections constituted the base of the next calculation and so on. Lorenz's first observation was that his artificial weather forecast very much resembled a natural weather forecast. He then concluded that, in the case of meteorology, randomness was not related to the complexity of the system. His second observation was that a simple rounding of 1 for 1,000 in his model resulted in considerable variations during the forecast period: in order to carry out his simulations over a longer period of time than he did in a past experiment, he recopied the system at the half-way point. To his great surprise, the second simulation slowly diverged from the first until at a certain moment it no longer resembled it at all.

Sensitivity to initial conditions

The sensitivity to initial conditions explains the unpredictability of a system at a given point in time. Quantitative models cannot identify and explain the exact process at play for two main reasons:

- *They cannot calculate the past initial point.* Referring to the past cannot produce estimation. This concept is related to that of entropy, which in the field of thermodynamics means that all energy transformations are accompanied by a loss which is irreversible. Dacunha-Castelle (1996) noted,

-
  "Entropy is thought of in terms of the quantity of randomness brought by the experiment, information is part of this in terms of the quantity of uncertainty suppressed by the observation of the results of the experiment" (pp. 207–208).

- *They cannot calculate the present point in time.* Not all information is immediately integrated in prices; consequently, we cannot measure the exact reactions of the market players. However, they can sometimes evaluate the

Y. Hayek Kobeissi, *Multifractal Financial Markets*, SpringerBriefs in Finance, DOI: 10.1007/978-1-4614-4490-9, © The Author(s) 2013

margin within which prices will fluctuate. But, since prices vary according to investors' perceptions of the value of an asset, this margin changes with time. Also, new macroeconomic or microeconomic information can, depending on their relevance, change the trading range of fluctuations

[ii]**As per Morse** (1980) **and Groth** (1979) **analysis**, the autocorrelation structure of f(t) is dependent on the amount of undisseminated (old as well as new) information as in the below equation:

$$\mathbf{R(t) = R(t) + f[I(t), t]}$$

R(t) is a Poisson-distributed mean jump process,

f (t) has a covariance matrix with autocorrelation terms that varies over time in a direct response to the amount of new information arriving into the market. The covariance matrix is nonstationary over time as a result of the sporadic jump process in the sequential information process and the varying finite speed of information dissemination.

[iii]**Equilibrium and disequilibrium conditions in such a market are derived from Nicolis and Prigogine** (Nicolis1977):

$$g_t - g_{t-1} = D_t - I_t$$

Dt is the amount of information that the market can disseminate for some time unit **t** (speed of information dissemination);

It is the amount of new information that has arrived in the market in a discrete sequential arrival process; and gt is the amount of information (in nits, bits, or Hartley) that has not been disseminated at any point in time **t**. It/Dt is the amount of time needed for the market to disseminate new information throughout the market at that particular time period, **t**.

The variable gt acts as a measure of disequilibrium, i.e., the greater the amount of undisseminated information, the greater the divergence from equilibrium. The variable gt is the stock of (undisseminated) information in the market and is defined by (Murphy 1965) in equation:

$$g_t = -(H * H_t),$$

H* is the maximum entropy and Ht is the actual entropy of the system. Whenever $D_t > (I_t - g_{t-1})$, then the actual entropy is maximized (H = H*) and there is no stock of information in the market (g = 0) since all information is disseminated. The value of g can be detected statistically by tests such as correlation analysis, runs tests, spectral analysis, and chi-square tests, since if there is undisseminated information, security returns will be dependent over time. However, as Pr. Nawrocki states, the setting of this limit so that it reflects the economic realities of transaction costs, taxes, and noisy information systems is difficult. Consequently, using g as an indicator is very limited and should be backed by other indicators.

[iv]**SINGLE BENCHMARK UPM / LPM UTILITY** (**Viole & Nawrocki** (2011a), 2011)

Our external benchmarks coincide directly with the various LPM and UPM levels of the utility and are not necessarily derived in a ubiquitous manner. Individuals are quite erratic in their rationale for selecting benchmark levels as evidenced by the reasons provided in historical utility experiments. Whether predetermined or path dependent, quantitatively or qualitatively derived, a loss of X will generate an LPM(n, h, x) to the investor to be then used in the function deriving a utility value for said loss. The investor will also have a benchmark to which the loss is compared. Typically, a benchmark for an equity investment is the broader S&P 500 index. The downside measure of the investment, LPM(n, 0, x) would then be compared to the LPM(n, 0, y), or the downside measure of the S&P 500. In this example, Y = S&P 500 and the target for the LPM calculations would be 0, identifying the nominal downside measures for both. Another benchmark, Z can also be introduced. Z often represents the risk-free rate or 3-month T-Bill for equity investments. So now you have LPM(n, 0, x), the investment, LPM(n, 0, y) the S&P 500, and LPM(n, 0, z) the 3-month T-Bill. The same benchmarks can be used for UPM calculations, or if the individual wishes to compare their upside measurements to a different set of benchmarks the function can accommodate it. The target for all benchmarks does not have to be 0. For instance, the LPMs in the prior example can be computed with the nonstationary 3-month T-Bill as the target and the downside measurements compared from that point, instead of a stationary target. There are no limits to the number of benchmarks or constraints on target selection; this is left entirely up to the individual.

[v]EVT explained

A straightforward definition states that, if you generate N data sets from the same distribution and create a new data set that includes the maximum values from these N data sets, the resulting data set can only be described by one of the three models—specifically, the Gumbel, Fréchet, and Weibull distributions. The Generalized Extreme Value (GEV) distribution combines the three-parameter model. It has the following probability density function (PDF):

$$f(x) = \begin{cases} \frac{1}{\sigma}\exp(-(1+kz)^{-1/k})(1+kz)^{-1-1/k} & k \neq 0 \\ \frac{1}{\sigma}\exp(-z - \exp(-z)) & k = 0 \end{cases}$$

Where $z = (x-\mu)/\sigma$ and k, σ, μ are the shape, scale, and location parameters respectively. The scale must be positive (*sigma* > 0), the shape and location can take any real value. The range of definition of the GEV distribution depends on k:

$$1 + k\frac{(x-\mu)}{\sigma} > 0 \quad \text{for } k \neq 0$$
$$-\infty < x < +\infty \quad \text{for } k = 0$$

Various values of the shape parameter will yield the extreme value type I, II, and III distributions. Specifically, the three cases k = 0, k > 0, and k < 0 correspond to the Gumbel, Fréchet, and "reversed" Weibull distributions. The reversed Weibull distribution is a quite rarely used model bounded on the upper side. When fitting the GEV distribution to sample data, the sign of the shape parameter k will usually indicate which one of the three models best describes the random process you are dealing with.

Could EVT have saved the space shuttle, Challenger?

The explosion of the space shuttle, Challenger, on 28 January, 1983 is an important example with respect to risk management as it was the consequence of an extreme event: the exceptionally low temperature (15 ° F lower than the next coldest previous launch) the night before launching ultimately led to failure of the o-rings which caused the disaster. Using standard EVT-analysis, one could have spotted that the shuttle should not have been launched at such cold temperature, despite having no measurements at such low temperature. Meanwhile, the LTCM's worst-case scenario estimated only a 20 % loss instead of the 60 % they actually incurred once things started to go wrong. EVT might have helped those (Chavez-Demoulin & Roehrl 2004).

[vi]EVT limits

Julia Schaumburg (Schaumburg 2010) findings in "Predicting extreme VaR: Nonparametric quantile regression with refinements from extreme value theory" states that "Modeling 1 % VaR a monotonized double kernel local linear of Yu and Jones (1998) estimator clearly outperforms competing models on the 1 % VaR level. Modeling 0.1 % VaR for extreme quantiles, usually very few data points are available, so that fully nonparametric regression does not yield reliable estimates. By refining nonparametric quantile regression methods with extreme value theory (EVT), we are able to model extreme quantiles (0.1 %) accurately."

L. Liubeckiene contribution to the existing Entropy Pooling theory (Liubeckiene 2009) is "the idea of applying Entropy Pooling technique to express the views on the joint default distribution to allow the stress-testing of the loan portfolio. It is useful for a default distribution as it works with non-normal markets and it handles views on nonlinear combinations of risk factors which affect returns directly or indirectly through co dependence. Specifically, we can investigate what happens when the dependence between defaults increases, throughout the whole distribution or just the low end of it. Stress-testing the tail co dependence means picking out the rows in the tail end of the copula where bad events happen together and increasing their occurrence probabilities." *But by how much this can be done, is a difficulty we face when applying it.*

As for Meucci (2011) Copula-marginal algorithm new CMA technique, it enables us to extract the copulas and the marginals from arbitrary joint distributions; to perform arbitrary transformations of those extracted copulas; and then to glue those transformed copulas back with another set of arbitrary marginal distributions. CMA can generate scenarios for many more copulas than the few parametric families used in the traditional approach. For instance, it includes large-dimensional, downside-only panic copulas which can be coupled with, say, extreme value theory marginals for portfolio stress-testing. An additional benefit of CMA is that it does not assume that all the scenarios have equal probabilities. *So what about correlations now?*

Printed by Publishers' Graphics LLC
LSI20121221.19.32.266